Solar Energy Pocket Reference

ISES

International
Solar Energy
Society

Villa Tannheim
Wiesentalstr. 50
79115 Freiburg
Germany
www.ises.org

Christopher L. Martin, Ph.D.
D. Yogi Goswami, Ph.D.

T0315472

First published 2005 by Earthscan

2 Park Square, Milton Park, Abingdon, Oxfordshire OX14 4RN
52 Vanderbilt Avenue, New York, NY 10017

Routledge is an imprint of the Taylor & Francis Group, an informa business

First issued in hardback 2019

Earthscan Publishes in association with the International Institute for Environment and Development.

ISBN 13: 978-1-84407-306-1 (pbk)
ISBN 13: 978-1-138-46873-3 (hbk)

Preface

The ISES *Solar Energy Pocket Reference* contains a lot of useful information and data needed by solar energy professionals, researchers, designers, technicians and those generally interested in solar energy systems and applications. The pocket book contains data, equations, figures and plots for worldwide solar radiation, solar angles, sunpath and shading, material properties, solar thermal collectors, energy storage, photovoltaic systems design and applications, solar water and space heating, daylighting design data, solar dryer configurations, life cycle costing and units and conversion factors. This information is gathered from books and other published literature and reviewed by a number of experts including scientists, engineers, and architects. We wish to express our deep gratitude to all the reviewers of this pocket book.

This pocket reference book is the first in a series of pocket books to be published by ISES, such as Wind Energy and Biomass, etc. We hope this series will serve the practitioners of the various renewable energy fields.

D. Yogi Goswami, Ph.D., P.E.
President, ISES

Contents

Sun/Earth Geometric Relationship

Annual motion of earth around the sun.

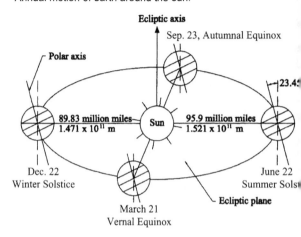

Location of the tropics.

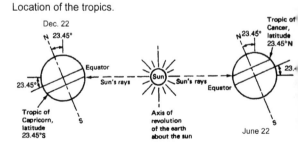

Solar Angle Formulations

titude angle, L, the angle between a line from the center of
e earth to the site of interest and the equatorial plane. Values
rth of the equator are positive and those south are negative.

lar declination, δ_s, the angle between the earth-sun line
rough their centers) and the plane through the equator.
clinations are positive in the northern hemisphere and
gative in the southern hemisphere.

$$\delta_s = 23.45° \sin\left(\frac{360(284 + n)}{365°}\right)$$

here n is the day number (Jan. 1 = 1, Jan 2 = 2, etc.)

ur angle, h_s, based on the nominal time of 24 hours for the
n to travel 360°, or 15° per hour. When the sun is due south
orthern hemisphere, due north for southern hemisphere), the
ur angle is 0, morning values are negative, and afternoon
ues are positive.

$$h_s = 15° \times (\text{hours from local solar noon})$$

$$= \frac{(\text{minutes from local solar noon})}{4\,(\text{min/deg})}$$

lar altitude angle, α, is the angle between a line collinear
th the sun's rays and the horizontal plane.

$$\sin\alpha = \sin L \sin\delta_s + \cos L \cos h_s$$

lar azimuth angle, a_s, is the angle between the projection of
e earth-sun line on the horizontal plane and the due south
ection (northern hemisphere) or due north (southern
misphere).

$$a_s = \sin^{-1}\left(\frac{\cos\delta_s \sin h_s}{\cos\alpha}\right)$$

te: $a_s > 90°$ that is $a_s\,(\text{am/pm}) = \mp\left(180° - |a_s|\right)$ when:

$$\begin{cases} L \le \delta_s \\ L > \delta_s \text{ and solar time earlier than } t_E \text{ or later than } t_W \end{cases}$$

$$t_E / t_W = 12 : 00 \text{ noon} \mp \left(\cos^{-1} \left[\tan \delta_s / \tan L \right] \right) / 15 \, (\text{deg}/\text{h})$$

Solar Time

Relation between local solar time and local standard time (LST)

$$\text{Solar Time} = \text{LST} + \text{ET} + \left(I_{st} - I_{local} \right) \times 4 \frac{min}{deg}$$

Where I_{st} is the standard time meridian, page 5, and I_{local} is the local longitude value. The equation of time, ET, can be gathered from tabular values, page 4, or computed with:

$$\text{ET} (\text{minutes}) = 9.87 \sin 2B - 7.53 \cos B - 1.5 \sin B$$

where: $$B = \frac{360 (n - 81)}{364°}$$

At local solar noon:

$$h_s = 0 \qquad\qquad \alpha = 90° - \left| L - \delta_s \right| \qquad\qquad a_s = 0$$

Sunrise and Sunset Times

At sunrise or sunset the center of the sun is located at the eastern or western horizon, respectively. By definition, the altitude angle is then: $\alpha = 0$. The local solar time for sunrise/sunset can be computed with:

$$\text{Sunrise} / \text{Sunset} = 12 : 00 \text{ noon} \mp \left[\frac{\left(\cos^{-1} \left(-\tan L \times \tan \delta_s \right) \right)}{15 \, (\text{deg}/\text{hour})} \right]$$

For the tip of the sun at the horizon (apparent sunrise/sunset), subtract/add approximately 4 minutes from the sunrise/sunset times.

3

lues for Declination, δs, and Equation of Time, ET

	Declination		Equation of Time			Declination		Equation of Time	
Date	Deg	Min	Min	Sec	Date	Deg	Min	Min	Sec
n. 1	−23	4	−3	14	Feb. 1	−17	19	−13	34
5	22	42	5	6	5	16	10	14	2
9	22	13	6	50	9	14	55	14	17
13	21	37	8	27	13	13	37	14	20
17	20	54	9	54	17	12	15	14	10
21	20	5	11	10	21	10	50	13	50
25	19	9	12	14	25	9	23	13	19
29	18	9	123	5					
r. 1	−7	53	−12	38	Apr. 1	+4	14	−4	12
5	6	21	11	48	5	5	46	3	1
9	5	48	10	51	9	7	17	1	52
13	3	14	9	49	13	8	46	−0	47
17	1	39	8	42	17	10	12	+0	13
21	−0	5	7	32	21	11	35	1	6
25	+1	30	6	20	25	12	56	1	53
29	3	4	5	7	29	14	13	2	33
y 1	+14	50	+2	50	June 1	+21	57	2	27
5	16	2	34	17	5	22	28	1	49
9	17	9	3	35	9	22	52	1	6
13	18	11	3	44	13	23	10	+0	18
17	19	9	3	44	17	23	22	−0	33
21	20	2	3	24	21	23	27	1	25
25	20	49	3	16	25	23	25	2	17
29	21	30	2	51	29	23	17	3	7
y 1	+23	10	−3	31	Aug. 1	+18	14	−6	17
5	22	52	4	16	5	17	12	5	59
9	22	28	4	56	9	16	6	5	33
13	21	57	5	30	13	14	55	4	57
17	21	21	5	57	17	13	41	4	12
21	20	38	6	15	21	12	23	3	19
25	19	50	6	24	25	11	2	2	18
29	18	57	6	23	29	9	39	1	10
. 1	+8	35	−0	15	Oct. 1	−2	53	+10	1
5	7	7	+1	2	5	4	26	11	17
9	5	37	2	22	9	5	58	12	27
13	4	6	3	45	13	7	29	13	30
17	2	34	5	10	17	8	58	14	25
21	1	1	6	35	21	10	25	15	10
25	0	32	8	0	25	11	50	15	46
29	2	6	9	22	29	13	12	16	10
v. 1	−14	11	+16	21	Dec. 1	−21	41	11	16
5	15	27	16	23	5	22	16	9	43
9	16	38	16	12	9	22	45	8	1
13	17	45	15	47	13	23	6	6	12
17	18	48	15	10	17	23	20	4	47
21	19	45	14	18	21	23	26	2	19
25	20	36	13	15	25	23	25	+0	20
29	21	21	11	59	29	23	17	−1	39

*Since each year is 365.25 days long, the precise value of declination varies from year to year. *The American Ephemeris and Nautical Almanac*, published each year by the U.S. Government Printing Office, ntains precise values for each day of each year.

Worldwide standard time meridians

Relative to UTC	Description	Standard Meridian, I_{st} [°]
-11	Nome Time (US) Samoa Standard Time	165° W
-10	Hawaiian Standard Time (US) Tahiti Time	150° W
-9	Alaska Standard Time (US) Yukon Standard Time	135° W
-8	US Pacific Standard Time	120° W
-7	US Mountain Standard Time	105° W
-6	US Central Standard Time Mexico Time	90° W
-5	US Eastern Standard Time Colombia Time	75° W
-4	Atlantic Standard Time Bolivia Time	60° W
-3	Eastern Brazil Standard Time Argentina Time	45° W
-2	Greenland Eastern Std. Time Fernando de Noronha Time (Brazil)	30° W
-1	Azores Time Cape Verde Time	15° W
0	Greenwich Mean Time Coordinated Universal Time Western Europe Time	0°
+1	Central Europe Time Middle European Time	15° E
+2	Eastern Europe Time Kaliningrad Time (Russia)	30° E
+3	Moscow Time (Russia) Baghdad Time	45° E
+4	Volga Time (Russia) Gulf Standard Time	60° E
+5	Yekaterinburg Time (Russia) Pakistan Time	75° E
+5.5	Indian Standard Time	82.5° E
+6	Novosibirsk Time (Russia) Bangladesh Time	90° E
+7	Krasnoyarsk Time (Russia) Java Time	105° E
+8	Irkutsk Time (Russia) China Coast Time	120° E
+9	Yakutsk Time (Russia) Japan Standard Time	135° E
+10	Vladivostok Time (Russia) Guam Standard Time	150° E
+11	Magadan Time (Russia) Solomon Islands Time	165° E
±12	Kamchatka Time (Russia) New Zealand Standard Time	180°

un Path Diagrams

ocedure to determine the solar altitude and azimuth angles for
given latitude, time of year, and time of day:

1. **Transition the time of interest, local standard time
   (LST), to solar time.** See procedure page 3.

2. **Determine the declination angle based on time of
   year.** Use formulation page 2, or table page 4.

3. **Read the solar altitude and azimuth angles from
   the appropriate sun path diagram.** Diagrams are
   chosen based on latitude; linear interpolations are
   used for latitudes not covered. For values at southern
   latitudes change the sign of the solar declination.

presentative path diagram illustrating the determination of
ar position.

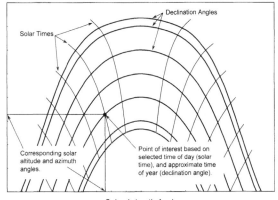

Solar Azimuth Angle

te: The sign convention for the declination angle, δ_s, is for
northern latitudes. To use the diagrams for southern
latitudes, reverse the sign of the declination angle.

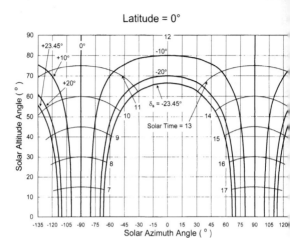

Latitude = 0°

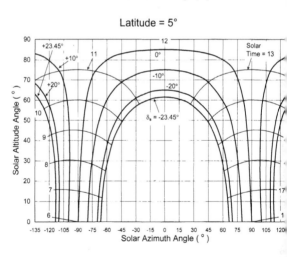

Latitude = 5°

Note: The sign convention for the declination angle, δ_s, is f
northern latitudes. To use the diagrams for southern
latitudes, reverse the sign of the declination angle.

Latitude = 10°

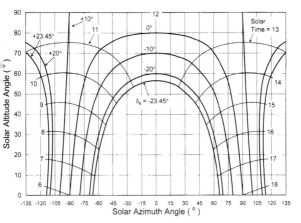

Latitude = 15°

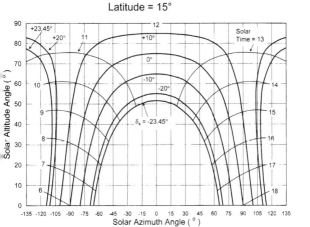

Note: The sign convention for the declination angle, δ_s, is for northern latitudes. To use the diagrams for southern latitudes, reverse the sign of the declination angle.

8

Latitude = 20°

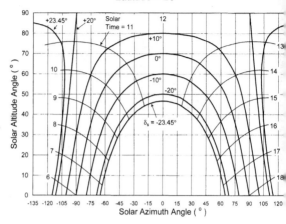

Latitude = 25°

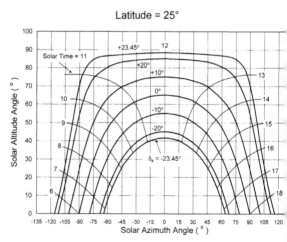

Note: The sign convention for the declination angle, δ_s, is for northern latitudes. To use the diagrams for southern latitudes, reverse the sign of the declination angle.

9

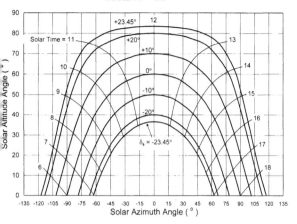

Latitude = 30°

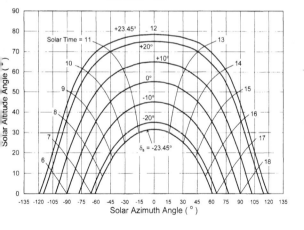

Latitude = 35°

te: The sign convention for the declination angle, δ_s, is for northern latitudes. To use the diagrams for southern latitudes, reverse the sign of the declination angle.

10

Latitude = 40°

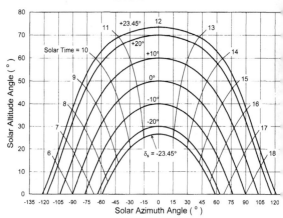

Latitude = 45°

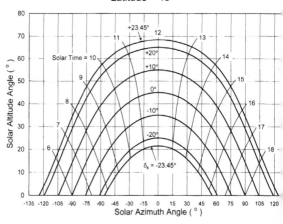

Note: The sign convention for the declination angle, δ_s, is fo[r] northern latitudes. To use the diagrams for souther[n] latitudes, reverse the sign of the declination angle.

11

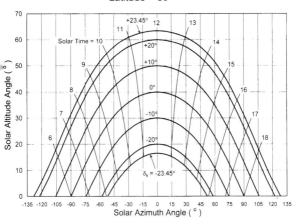

Latitude = 50°

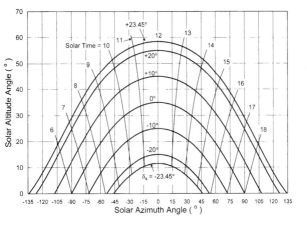

Latitude = 55°

ote: The sign convention for the declination angle, δ_s, is for northern latitudes. To use the diagrams for southern latitudes, reverse the sign of the declination angle.

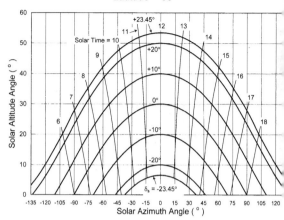

Latitude = 60°

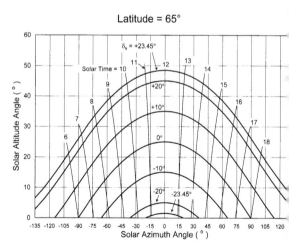

Latitude = 65°

Note: The sign convention for the declination angle, δ_s, is for northern latitudes. To use the diagrams for southern latitudes, reverse the sign of the declination angle.

13

Overhang and Fin Shading Calculation

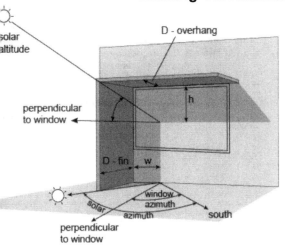

Reprinted from [3]

For overhang shading:

$$= \frac{D \times \tan\left(\text{solar altitude}\right)}{\cos\left(\text{solar azimuth} - \text{window azimuth}\right)} *$$

For fin shading:

$$= D \times \tan\left(\text{solar azimuth} - \text{window azimuth}\right) *$$

Solar altitude and azimuth angles can be determined with the sun path diagrams, beginning on page 6.

Observe proper signs for both the solar and window azimuth values. Convention is that angles east of the equator vector are negative while those west are positive.

Obstruction Shading

Procedure for estimating shading caused by adjacent structure▮

1. **Define the solid angle presented by the obstruction in** **terms of altitude, α_{obs}, and azimuth, a_{obs}, angles for th▮** **point of interest.** Choose characteristic points to describe the profile ▮ the obstruction and determine their angular coordinates; refer to the geometr▮ of the figures below and on the next page. By measuring or estimating the distances and heights in question, standard trigonometric formulas (next pag▮ can be used to determine the angular values.

2. **Locate the solid angle with respect to the path of the** **sun.** Refer to the appropriate sun path diagram, based on the site latitude, and locate the solid angle on the diagram by plotting the α_{obs} and a_{obs} values the solar altitude and solar azimuth axes, respectively. An example of this positioning is shown on the next page.

3. **Determine dates and times of shading.** On the sun path diagram, where the solid angle overlaps the declination lines, shading of the point of interest will occur. The dates can be estimated by reading the declination values that intersect the solid angle, and then matching the declination to the time of year, table page 4. In a similar manner, the hours ▮ shading for a particular declination value, can be estimated by reading the s▮ time corresponding to the intersection of the declination of interest and the s▮ angle presented by the obstruction.

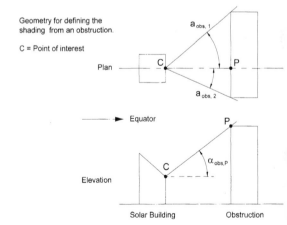

Geometry for defining the
shading from an obstruction.

C = Point of interest

Plan

$a_{obs, 1}$

C

P

$a_{obs, 2}$

Equator

P

C

$\alpha_{obs,P}$

Elevation

Solar Building   Obstruction

Obstruction Shading

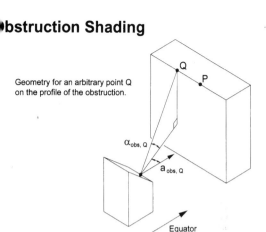

Geometry for an arbitrary point Q on the profile of the obstruction.

Example overlay of an obstruction profile on a sun path diagram.

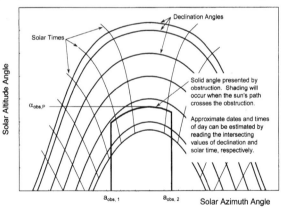

$$\theta = \sin^{-1}\left(\frac{b}{c}\right) \quad \theta = \tan^{-1}\left(\frac{b}{a}\right)$$

$$\theta = \cos^{-1}\left(\frac{a}{c}\right) \quad c^2 = a^2 + b^2$$

16

Annual average horizontal surface insolation

July 1983 – June 1993
Source: Surface meteorology and solar energy, National Aeronautics and Space
Administration, USA; http://eosweb.larc.nasa.gov/sse

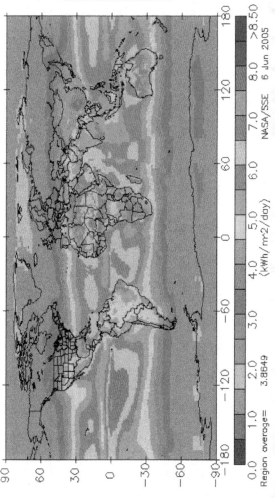

17

Worldwide, Horizontal Surface
Solar Radiation Data, [MJ/m²-day]

Position	Lat	Long	Jan.	Feb.	Mar.	Apr.	May	June	July	Aug.	Sep.	Oct.	Nov.	Dec.
Argentina														
Buenos Aires	34.58 S	58.48 W	24.86	21.75	18.56	11.75	8.71	7.15	7.82	8.75	14.49	16.66	24.90	21.93
Australia														
Adelaide	34.93 S	138.52 E	20.99	17.50	20.15	18.27	17.98	—	18.81	19.64	20.11	20.88	20.57	20.72
Brisbane	27.43 S	153.08 E	25.36	22.22	13.25	16.61	12.23	11.52	9.70	15.10	17.61	19.89	—	—
Canberra	35.30 S	148.18 E	28.20	24.68	20.56	14.89	10.29	6.62	—	12.33	16.88	24.06	26.00	25.77
Darwin	12.47 S	130.83 E	26.92	23.40	18.13	13.62	9.30	7.89	9.41	11.15	14.85	18.87	23.43	22.34
Hobart	42.88 S	147.32 E	—	—	—	10.09	7.26	6.04	5.72	9.21	13.54	18.12	—	—
Laverton	37.85 S	114.08 E	22.96	20.42	15.59	13.40	7.48	6.10	6.54	10.43	13.24	18.76	—	—
Sydney	33.87 S	151.20 E	21.09	21.75	17.63	13.63	9.78	8.79	7.62	12.84	16.93	22.10	—	—
Austria														
Wien	48.20 N	16.57 E	3.54	7.10	8.05	14.72	16.79	20.87	19.89	17.27	12.55	8.45	3.51	2.82
Innsbruck	47.27 N	11.38 E	5.57	9.28	10.15	15.96	14.57	17.65	18.35	17.26	12.98	9.08	4.28	3.50
Barbados														
Husbands	13.15 N	59.62 W	19.11	20.23	—	21.80	19.84	20.86	21.55	22.14	—	—	18.30	16.56
Belgium														
Ostende	51.23 N	2.92 E	2.82	5.75	9.93	15.18	16.74	16.93	18.21	18.29	11.71	6.15	2.69	1.97
Melle	50.98 N	3.83 E	2.40	4.66	8.41	13.55	14.23	13.28	15.71	15.61	10.63	5.82	2.40	1.59
Brunei														
Brunei	4.98 N	114.93 E	19.46	20.12	22.71	20.54	19.74	18.31	19.38	20.08	20.83	17.51	17.39	18.12
Bulgaria														
Chirpan	42.20 N	25.33 E	6.72	6.79	8.54	13.27	17.25	17.39	19.85	14.61	12.53	8.52	5.08	5.09
Sofia	42.65 N	23.38 E	4.05	6.23	7.93	9.36	12.98	19.73	19.40	17.70	14.71	6.44	—	3.14
Canada														
Montreal	45.47 N	73.75 E	4.74	8.33	11.84	10.55	15.05	22.44	21.08	18.67	14.83	9.18	4.04	4.01
Ottawa	45.32 N	75.67 E	5.34	9.59	13.33	13.98	20.18	20.34	19.46	17.88	13.84	7.38	4.64	5.04
Toronto	43.67 N	79.38 E	4.79	8.15	11.96	14.00	18.16	24.35	23.38	—	15.89	9.40	4.72	3.79
Vancouver	49.18 N	123.17 E	3.73	4.81	12.14	16.41	20.65	24.04	22.87	19.08	12.77	7.39	4.29	1.53
Chile														
Pascua	27.17 S	109.43 W	19.64	16.65	—	11.12	9.52	8.81	10.90	12.29	17.19	20.51	21.20	22.44
Santiago	33.45 S	70.70 W	18.61	16.33	13.44	8.32	5.07	3.66	3.35	5.65	8.15	13.62	20.14	23.88

Worldwide, Horizontal Surface Solar Radiation Data, [MJ/m²-day]

Position	Lat	Long	Jan.	Feb.	Mar.	Apr.	May	June	July	Aug.	Sep.	Oct.	Nov.	Dec.
China														
Beijing	39.93 N	116.28 W	7.73	10.59	13.87	17.93	20.18	18.65	15.64	16.61	15.52	11.29	7.25	6.89
Guangzhou	23.13 N	113.32 E	11.01	6.32	4.04	7.89	10.53	12.48	16.14	16.02	15.03	15.79	11.55	9.10
Harbin	45.75 N	126.77 E	5.15	9.54	17.55	20.51	20.33	17.83	19.18	16.09	14.50	14.50	10.50	6.98
Kunming	25.02 N	102.68 E	9.92	11.26	14.38	18.00	18.53	17.37	11.95	18.47	15.94	12.45	11.96	13.62
Lanzhou	36.05 N	103.88 E	7.30	12.47	10.62	18.91	17.40	20.40	20.23	17.37	13.23	10.21	8.22	6.43
Shanghai	31.17 N	121.43 E	7.44	10.31	11.78	14.36	14.23	16.79	14.63	11.85	15.96	12.03	7.73	8.70
Columbia														
Bogota	4.70 N	74.13 W	17.89		19.37	16.58	14.86		15.42	18.20	17.05	14.58	14.20	16.66
Cuba														
Havana	23.17 N	82.35 W		14.70	18.94	20.95	22.63	18.83	21.40	20.19	16.84	16.98	13.19	13.81
Czech														
Kucharovice	48.88 N	16.08 E	3.03	5.85	9.88	14.06	20.84	19.24	21.18	19.41	13.61	6.11	3.47	2.12
Churanov	49.07 N	13.62 E	2.89	5.82	9.24	13.18	21.32	15.58	20.51	19.49	12.84	5.68	3.36	2.99
Hradec Kralov	50.25 N	15.85 E	3.51	5.94	10.58	15.95	20.42	18.43	17.17	17.92	11.86	6.27	2.45	1.89
Denmark														
Copenhagen	55.67 N	12.30 E	1.83	3.32	7.09	11.12	21.39	24.93		13.92	10.10	5.20	2.81	1.23
Egypt														
Cairo	30.08 N	31.28 E	10.06	12.96	18.49	23.04	21.91	26.07	25.16	23.09	21.01		11.74	9.85
Mersa Maaruh	31.33 N	27.22 E	8.38	11.92	18.47	24.27	24.17		26.67	26.27	21.92	18.28	11.71	8.76
Ethiopia														
Addis Ababa	8.98 N	38.80 E		11.39		12.01				6.33	9.35	11.71	11.69	11.50
Fiji														
Nandi	17.75 S	177.45 E	20.82	20.65	20.25	18.81	15.68	14.18	15.08	16.71	19.37	20.11	21.78	25.09
Suva	48.05 S	178.57 E	20.37	17.74	16.22	13.82	10.81	12.48	11.40			18.49	19.96	20.99
Finland														
Helsinki	60.32 N	24.97 E	1.13	2.94	5.59	11.52	17.60	16.81	20.66	15.44	8.44	3.31	0.97	0.63
France														
Agen	44.18 N	0.60 E	4.83	7.40	10.69	17.12	19.25	20.42	21.63	20.64	15.56	8.41	5.09	5.01
Nice	43.65 N	7.20 E	6.83		11.37	17.79	20.74	24.10	24.85	24.86	15.04	10.99	7.08	6.73
Paris	48.97 N	2.45 E	2.62	5.08	7.21	12.90	14.84	13.04	15.54	16.30	10.17	5.61	3.14	2.20

Worldwide, Horizontal Surface Solar Radiation Data, [MJ/m²-day]

Position	Lat	Long	Jan	Feb	Mar	Apr	May	June	July	Aug	Sep	Oct	Nov	Dec
Germany														
Bonn	50.70 N	7.15 E	2.94	5.82	8.01	14.27	15.67	14.41	18.57	17.80	11.70	6.15	3.42	1.90
Nuremberg	53.33 N	13.20 E	3.23	6.92	9.08	15.69	15.71	18.21	21.14	17.98	12.43	8.15	2.79	2.51
Bremen	53.05 N	8.80 E	2.36	4.93	8.53	14.52	14.94	14.52	19.40	15.02	10.48	6.27	2.80	1.66
Hamburg	53.63 N	10.00 E	1.97	3.96	7.59	12.32	14.11	12.69	19.00	14.11	10.29	6.45	2.33	1.43
Stuttgart	48.83 N	9.20 E	3.59	7.18	9.22	15.81	17.72	17.44	22.21	19.87	12.36	7.81	3.19	2.54
Ghana														
Bole	9.03 N	2.48 W	18.29	19.76	19.71	19.15	16.61	—	—	13.68	16.29	17.27	17.33	15.93
Accra	5.60 N	0.17 W	14.82	16.26	18.27	16.73	18.15	13.96	13.86	13.49	15.32	19.14	18.16	14.23
Great Britain														
Belfast	54.65 N	6.22 W	2.00	3.60	6.85	12.00	15.41	15.09	15.46	13.56	11.49	4.63	2.34	1.24
Jersey	49.22 N	2.20 W	2.76	5.65	9.51	14.98	18.51	17.83	18.14	18.62	12.98	6.16	3.26	2.83
London	51.52 N	0.12 W	2.24	3.87	7.40	12.01	12.38	13.24	16.59	16.23	12.59	5.67	2.87	1.97
Greece														
Athens	37.97 N	23.72 E	9.11	10.94	15.70	20.91	23.85	25.48	24.21	23.08	19.03	13.29	5.98	6.64
Sikkiwa	37.98 N	22.73 E	7.60	8.16	11.99	21.06	22.62	24.32	23.56	21.73	17.30	11.75	9.45	6.35
Guadeloupe														
Le Raizet	16.27 N	61.52 W	14.88	18.10	20.55	19.69	20.26	20.65	20.65	20.24	18.47	17.79	13.49	14.38
Guyana														
Cayenne	4.83 N	52.37 W	14.46	14.67	16.28	17.57	—	14.92	17.42	18.24	20.52	—	22.69	17.04
Hong Kong														
King's Park	22.32 N	114.17 W	12.34	7.39	6.94	9.50	11.38	13.60	16.70	17.06	15.91	16.52	14.19	10.00
Hungary														
Budapest	47.43 N	19.18 E	2.61	7.46	11.14	14.46	20.69	19.47	21.46	19.72	12.88	7.96	2.95	2.47
Iceland														
Reykjavik	64.13 N	21.90 W	0.52	2.02	6.25	11.77	13.07	14.58	16.83	11.35	9.70	3.18	1.00	0.65
India														
Bombay	19.12 N	72.85 E	18.44	21.00	22.72	24.52	24.86	19.75	15.84	16.00	18.19	20.38	19.18	17.81
Calcutta	22.53 N	88.33 E	15.69	18.34	20.09	22.34	22.37	17.55	17.07	16.55	16.52	16.90	16.35	15.00
Madras	13.00 N	80.18 E	19.09	22.71	25.14	24.88	23.89	—	18.22	19.68	19.81	16.41	14.76	15.79

Worldwide, Horizontal Surface Solar Radiation Data, [MJ/m²-day]

Position	Lat	Long	Jan.	Feb.	Mar.	Apr.	May	June	July	Aug.	Sep.	Oct.	Nov.	Dec.
India														
Nagpur	21.10 N	79.05 E	18.08	21.01	22.25	24.08	24.79	19.84	15.58	15.47	17.66	20.10	18.98	17.33
New Delhi	28.58 N	77.20 E	14.62	18.25	20.15	23.40	23.80	19.16	20.20	19.89	20.08	19.74	16.95	14.22
Ireland														
Dublin	53.43 N	6.25 W	2.51	4.75	7.48	11.06	17.46	19.11	15.64	13.89	9.65	5.77	2.93	—
Israel														
Jerusalem	31.78 N	35.22 E	10.79	13.01	18.08	23.79	29.10	31.54	31.83	28.79	25.19	20.26	12.61	10.71
Italy														
Milan	45.43 N	9.28 E	—	6.48	10.09	13.17	17.55	16.32	18.60	16.86	11.64	5.40	3.52	2.41
Rome	41.80 N	12.55 E	—	9.75	13.38	15.82	15.82	18.89	22.27	21.53	16.08	8.27	6.41	4.49
Japan														
Fukuoka	33.58 N	130.38 E	8.11	8.72	10.95	13.97	14.36	12.81	13.84	16.75	13.92	11.86	10.05	7.30
Tateno	36.05 N	140.13 E	9.06	12.17	11.00	15.78	16.52	15.26	—	—	—	9.60	8.55	8.26
Yonago	35.43 N	133.35 E	6.25	7.16	10.87	17.30	16.72	15.44	17.06	19.93	12.41	10.82	7.50	5.51
Kenya														
Mombasa	4.03 S	39.62 E	22.30	22.17	22.74	18.49	18.31	17.41	—	18.12	21.03	22.97	21.87	21.25
Nairobi	1.32 S	36.92 E	—	24.10	21.20	18.65	14.83	15.00	13.44	14.12	19.14	19.38	16.90	18.27
Lithuania														
Kaunas	54.88 N	23.88 E	1.89	4.43	7.40	12.97	18.88	18.74	21.41	15.79	10.40	5.64	1.80	1.10
Madagascar														
Antananarivo	18.80 S	47.48 E	15.94	13.18	13.07	11.53	9.25	8.21	9.32	—	—	16.43	15.19	15.62
Malaysia														
Kualalumpur	3.12 N	101.55 E	15.36	17.67	18.48	16.87	15.67	16.24	15.32	15.89	14.62	14.13	13.54	11.53
Piang	5.30 N	100.27 E	19.47	21.35	23.24	20.52	18.63	19.32	17.17	16.96	15.93	16.01	18.35	17.37
Martinique														
Le Lamentin	14.60 N	61.00 W	17.76	20.07	22.53	21.95	22.42	21.23	20.86	21.84	20.23	19.87	14.08	16.25
Mexico														
Chihuahua	28.63 N	106.08 W	14.80	—	—	—	26.94	26.28	24.01	24.22	20.25	19.55	10.57	15.79
Orizaba	20.58 N	99.20 E	19.40	23.07	27.44	27.15	26.04	25.05	—	27.33	21.06	17.85	15.48	12.93

Worldwide, Horizontal Surface
Solar Radiation Data, [MJ/m²-day]

Position	Lat	Long	Jan.	Feb.	Mar.	Apr.	May	June	July	Aug.	Sep.	Oct.	Nov.	Dec.
Mongolia														
Ulan Bator	47.93 N	106.98 E	6.28	9.22	14.34	18.18	20.50	19.34	16.34	16.65	14.08	11.36	7.19	5.35
Ulinastai	47.75 N	96.85 E	6.43	10.71	14.83	20.32	23.86	20.46	21.66	17.81	15.97	10.92	7.32	5.08
Morocco														
Casablanca	33.57 N	7.67 E	11.46	12.70	15.93	21.25	24.45	25.27	25.53	23.60	19.97	14.68	11.61	9.03
Mozambique														
Maputo	25.97 S	32.60 E	26.35	23.16	19.33	20.54	16.33	14.17	—	—	—	22.55	25.48	26.19
Netherlands														
Maastricht	50.92 N	5.78 E	3.20	5.43	8.48	14.82	14.97	14.32	18.40	17.51	11.65	6.51	3.01	1.72
New Caledonia														
Koumac	20.57 S	164.28 E	24.89	21.15	16.96	18.98	15.67	14.55	15.75	17.62	22.48	15.83	27.53	26.91
New Zealand														
Wilmington	41.28 S	174.77 E	22.59	19.67	14.91	9.52	6.97	4.37	5.74	7.14	12.50	16.34	19.07	24.07
Christchurch	43.48 S	172.55 E	23.46	19.68	13.98	8.96	6.47	4.74	5.38	6.94	13.18	17.45	18.91	24.35
Nigeria														
Benin City	6.32 N	5.60 E	14.89	17.29	19.15	17.21	16.97	15.04	10.24	12.54	14.37	15.99	17.43	15.75
Norway														
Bergen	60.40 N	5.32 E	0.46	1.33	3.18	8.36	19.24	16.70	16.28	10.19	6.53	3.19	1.36	0.35
Oman														
Seeb	23.58 N	58.28 E	12.90	14.86	21.22	22.22	25.30	24.02	23.46	21.66	20.07	18.45	15.49	13.12
Salalah	17.03 N	54.08 E	16.52	16.92	18.49	20.65	21.46	16.92	8.52	11.41	17.14	18.62	16.42	—
Pakistan														
Karachi	24.90 N	67.13 E	13.84			19.69	20.31	16.62				—	12.94	11.07
Multan	30.20 N	71.43 E	12.29	15.86	18.33	22.35	22.57	21.65	20.31	20.44	20.57	15.91	12.68	10.00
Islamabad	33.62 N	73.10 E	10.38	12.42	16.98	22.65	—	25.49	20.64	18.91	14.20	15.30	10.64	8.30
Peru														
Puno	15.83 S	70.02 W	14.98	12.92	16.08	20.03	17.45	17.42	15.74	15.32	16.11	16.18	14.24	13.90
Poland														
Warszawa	52.28 N	20.97 E	1.73	3.83	7.81	10.53	19.22	17.11	20.18	15.00	10.65	4.95	2.39	1.68
Kolobrzeg	54.18 N	15.58 E	2.50	3.25	8.86	15.21	20.79	20.50	17.19	16.46	7.95	5.75	1.78	1.18

Worldwide, Horizontal Surface Solar Radiation Data, [MJ/m²-day]

Position	Lat	Long	Jan.	Feb.	Mar.	Apr.	May	June	July	Aug.	Sep.	Oct.	Nov.	Dec.
Portugal														
Evora	38.57 N	7.90 W	9.92	12.43	17.81	18.69	23.57	29.23	28.75	23.77	20.17	–	6.81	4.57
Lisbon	38.72 N	9.15 W	9.24	11.60	17.52	18.49	24.64	29.02	28.14	22.20	19.76	13.56	7.18	4.83
Romania														
Bucuresti	44.50 N	26.13 E	7.05	10.22	12.04	16.53	18.97	22.16	23.19	–	17.17	9.55	4.82	–
Constansa	44.22 N	28.63 E	5.62	9.28	14.31	20.59	23.23	25.80	27.98	24.22	16.91	11.89	6.19	5.10
Galati	45.50 N	28.02 E	6.09	9.33	14.31	17.75	21.77	22.74	25.55	19.70	14.05	11.26	6.32	5.38
Russia														
Alexandrovsko	60.38 N	77.87 E	1.34	4.17	9.16	17.05	21.83	21.34	20.26	13.05	10.16	4.68	1.71	0.68
Moscow	55.75 N	37.57 E	1.45	3.96	8.09	11.69	18.86	18.12	17.51	14.17	10.92	4.03	2.28	1.29
St. Petersburg	59.97 N	30.30 E	1.03	3.11	4.88	12.24	20.59	21.55	20.43	13.27	7.83	2.93	1.16	0.59
Verkhoyansk	67.55 N	133.38 E	0.21	2.25	7.61	15.96	19.64	–	–	14.12	7.59	3.51	0.54	–
St. Pierre & Miquelon														
St. Pierre	46.77 N	56.17 W	4.43	6.61	12.50	17.57	18.55	17.84	19.95	16.46	12.76	8.15	3.69	3.33
Singapore														
Singapore	1.37 N	103.98 E	19.08	20.94	20.75	18.20	14.89	15.22	13.92	16.66	16.51	15.82	13.81	12.67
South Korea														
Seoul	37.57 N	126.97 E	6.24	9.40	10.34	13.98	16.35	17.49	10.65	12.94	11.87	10.35	6.47	5.14
South Africa														
Cape Town	33.98 S	18.60 E	27.47	25.57	–	15.81	11.44	9.08	8.35	13.76	17.30	22.16	26.37	27.68
Port Elizabeth	33.98 S	25.60 E	27.22	22.06	19.01	15.29	11.79	11.13	10.73	13.97	18.52	23.09	23.15	27.26
Pretoria	25.73 S	28.18 E	26.06	22.43	20.52	16.09	15.67	13.67	15.19	18.65	21.62	21.75	24.82	23.43
Spain														
Madrid	40.45 N	3.72 W	7.73	10.53	15.35	21.74	22.81	22.05	26.27	22.90	18.89	10.21	8.69	5.56
Sudan														
Wad Madani	14.40 N	33.48 E	21.92	24.01	23.43	25.17	23.92	23.51	22.40	22.85	21.75	20.47	20.19	19.21
Elfasher	13.62 N	25.33 E	21.56	21.84	24.54	25.29	24.31	24.15	22.87	21.19	22.58	23.85	–	–
Shambat	15.67 N	32.53 E	23.90	27.38	–	27.45	23.21	26.15	23.55	25.46	24.05	23.51	23.82	22.53
Sweden														
Karlstad	59.37 N	13.47 E	1.26	3.13	5.02	14.01	19.90	16.70	20.92	14.14	10.42	3.98	1.62	0.9?

Position	Lat	Long	Jan.	Feb.	Mar.	Apr.	May	June	July	Aug.	Sep.	Oct.	Nov.	Dec.
Sweden														
Lund	55.72 N	13.22 E	1.97	3.47	6.66	12.48	17.83	13.38	18.74	14.99	10.39	5.45	1.82	1.21
Stockholm	59.35 N	18.07 E	1.32	2.69	4.75	13.21	15.58	14.79	20.52	14.48	10.50	4.04	1.19	0.83
Switzerland														
Geneva	46.25 N	6.13 E	2.56	7.21	9.46	17.07	20.98	19.78	22.38	20.50	13.62	8.44	3.31	2.87
Zurich	47.48 N	8.53 E	2.31	7.02	7.54	15.04	16.33	16.73	20.28	18.32	12.52	7.18	2.64	2.29
Thailand														
Bangkok	13.73 N	100.57 E	16.67	19.34	23.00	22.48	20.59	17.71	18.02	16.04	16.23	16.81	18.60	16.43
Trinidad & Tobago														
Crown Point	11.15 N	60.83 W	13.05	15.61	15.17	16.96	17.61	15.37	13.16	13.08	12.24	8.76	—	—
Tunisia														
Sidi Bouzid	36.87 N	10.35 E	7.88	10.38	13.20	17.98	25.12	26.68	27.43	24.33	18.87	12.11	9.37	6.72
Tunis	36.83 N	10.23 E	7.64	9.88	14.79	31.61	25.31	26.03	26.60	20.37	19.58	12.91	9.35	7.16
Ukraine														
Kiev	50.40 N	30.45 E	2.17	4.87	11.15	12.30	20.49	—	18.99	18.55	9.72	9.84	3.72	2.52
Uzbekistan														
Tashkent	41.27 N	69.27 E	7.27	10.81	15.93	23.60	25.21	29.53	28.50	26.68	20.76	13.25	8.61	4.59
Venezuela														
Caracas	10.50 N	66.88 W	14.25	13.56	16.30	15.56	15.69	15.56	16.28	17.11	17.04	15.14	14.74	13.50
St. Antonio	7.85 N	72.45 W	11.78	10.54	10.65	12.07	12.65	21.20	14.68	15.86	16.62	15.32	12.28	11.28
St. Fernando	7.90 N	67.42 W	14.92	16.82	16.89	—	—	14.09	13.78	14.42	14.86	15.27	14.25	13.11
Vietnam														
Hanoi	21.03 N	105.85 E	5.99	7.48	8.73	13.58	19.10	21.26	19.85	19.78	20.67	14.78	12.44	13.21

Worldwide, Horizontal Surface Solar Radiation Data, [MJ/m²-day]

Position	Lat	Long	Jan.	Feb.	Mar.	Apr.	May	June	July	Aug.	Sep.	Oct.	Nov.	Dec.
Yugoslavia														
Beograd	44.78 N	20.53 E	4.92	6.27	10.64	14.74	20.95	22.80	22.09	20.27	15.57	11.24	6.77	4.99
Kopaonik	43.28 N	20.80 E	7.03	10.93	14.75	12.78	13.54	20.43	22.48	—	20.14	11.61	6.26	4.64
Portoroz	45.52 N	13.57 E	5.11	7.84	13.75	17.30	23.66	22.31	25.14	21.34	13.40	8.98	6.04	3.92
Zambia														
Lusaka	15.42 S	28.32 W	16.10	18.02	20.24	19.84	17.11	16.37	19.45	20.72	21.68	23.83	23.85	20.52
Zimbabwe														
Bulawayo	20.15 S	28.62 N	20.03	22.11	21.03	18.09	17.15	15.36	16.46	19.49	21.55	23.44	25.08	23.46
Harare	17.83 S	31.02 N	19.38	19.00	19.22	17.67	18.35	16.10	14.55	17.87	21.47	23.98	19.92	21.88

(Source: Voeikov Main Geophysical Observatory, Russia: Internet address: http://wrdc-mgo.nrel.gov/html/get_data-ap.html)

Note: Data for 872 locations is available from these sources in 68 countries.

*Source for Canadian Data: Environment Canada: Internet address: http://www.ec.gc.ca/envhome.html.

Worldwide, Horizontal Surface Solar Radiation Data, [MJ/m²-day]

Position	Jan.	Feb.	Mar.	Apr.	May	June	July	Aug.	Sep.	Oct.	Nov.	Dec.	Average
Alabama													
Birmingham	9.20	11.92	15.67	19.65	21.58	22.37	21.24	20.21	17.15	14.42	10.22	8.40	16.01
Montgomery	9.54	12.49	16.24	20.33	22.37	23.17	21.80	20.56	17.72	14.99	10.90	8.97	16.58
Alaska													
Fairbanks	0.62	2.77	8.31	14.66	17.98	19.65	16.92	12.36	7.02	3.20	1.01	0.23	8.74
Anchorage	1.02	3.41	8.18	13.06	15.90	17.72	16.69	12.72	8.06	3.97	1.48	0.56	8.63
Nome	0.51	2.95	8.29	15.22	18.97	19.65	16.69	11.81	7.72	3.63	0.99	0.09	8.86
St. Paul Island	1.82	4.32	8.52	12.72	14.08	14.42	12.83	10.33	7.84	4.54	2.16	1.25	7.95
Yakutat	1.36	3.63	7.72	12.61	14.76	15.79	14.99	12.15	7.95	3.97	1.82	0.86	8.18
Arizona													
Phoenix	11.58	15.33	19.87	25.44	28.85	30.09	27.37	25.44	21.92	17.60	12.95	10.56	20.56
Tucson	12.38	15.90	20.21	25.44	28.39	29.30	25.44	24.08	21.58	17.94	13.63	11.24	20.44
Arkansas													
Little Rock	9.09	11.81	15.56	19.19	21.80	23.51	23.17	21.35	17.26	14.08	9.77	8.06	16.24
Fort Smith	9.31	12.15	15.67	19.31	21.69	23.39	23.85	24.46	17.26	13.97	9.88	8.29	16.35
California													
Bakersfield	8.29	11.92	16.69	22.15	26.57	28.96	28.73	26.01	21.35	15.90	10.33	7.61	18.74
Fresno	7.61	11.58	16.81	22.49	27.14	29.07	28.96	25.89	21.12	15.56	9.65	6.70	18.62
Long Beach	9.99	12.95	17.03	21.60	23.17	24.19	26.12	24.08	19.31	14.99	11.24	9.31	17.83
Sacramento	6.93	10.68	15.56	21.24	25.89	28.28	28.62	25.32	20.56	14.54	8.63	6.25	17.72
San Diego	11.02	13.97	17.72	21.92	22.49	23.28	24.98	23.51	19.53	15.79	12.26	10.22	18.06
San Francisco	7.72	10.68	15.22	20.44	24.08	25.78	26.46	23.39	19.31	13.97	8.97	7.04	16.92
Los Angeles	10.11	13.06	17.26	21.80	23.05	23.74	25.67	23.51	18.97	14.99	11.36	9.31	17.72
Santa Maria	10.22	13.29	17.49	22.26	25.10	26.57	26.91	24.42	20.10	15.67	11.47	9.54	18.62
Colorado													
Boulder	7.84	10.45	15.64	17.94	17.94	20.47	20.28	17.12	16.07	12.09	8.66	7.10	14.31
Colorado Springs	9.09	12.15	16.13	20.33	22.26	24.98	23.96	21.69	18.51	14.42	9.99	8.18	16.81
Connecticut													
Hartford	6.70	9.65	13.17	16.69	19.53	21.24	21.12	18.51	14.76	10.68	6.59	5.45	13.74

Worldwide, Horizontal Surface Solar Radiation Data, [MJ/m²-day]

Position	Jan.	Feb.	Mar.	Apr.	May	June	July	Aug.	Sep.	Oct.	Nov.	Dec.	Average
Delaware													
Wilmington	7.27	10.22	13.97	17.60	20.33	22.49	21.80	19.65	15.79	11.81	7.84	6.25	14.65
Florida													
Daytona Beach	11.24	13.85	17.94	22.15	23.17	22.03	21.69	20.44	17.72	14.99	12.15	10.33	17.38
Jacksonville	10.45	13.17	17.03	21.12	22.03	21.58	21.01	19.42	16.69	14.20	11.47	9.65	16.47
Tallahassee	10.33	13.29	16.92	21.24	22.49	22.03	20.90	19.65	17.72	15.56	11.92	9.77	16.81
Miami	12.72	15.22	18.51	21.58	21.46	20.10	21.10	20.10	17.60	15.67	13.17	11.81	17.38
Key West	13.17	16.01	19.65	22.71	22.83	22.03	22.03	21.01	18.74	16.47	13.85	15.79	18.40
Tampa	11.58	14.42	18.17	22.26	23.05	21.92	20.90	19.65	17.60	16.01	12.83	11.02	17.49
Georgia													
Athens	9.43	12.38	16.01	20.21	22.03	22.83	21.80	20.21	17.26	14.42	10.45	8.40	16.29
Atlanta	9.31	12.26	16.13	20.33	22.37	23.17	22.15	20.56	17.49	14.54	10.56	8.52	16.43
Columbus	9.77	12.72	16.47	20.67	22.37	22.83	21.58	20.33	17.60	14.99	11.02	9.09	16.62
Macon	9.54	12.61	16.35	20.56	22.37	22.83	21.58	20.21	17.60	14.88	10.90	8.86	16.50
Savanna	9.99	12.72	16.81	21.01	22.37	22.60	21.80	19.76	16.92	14.65	11.13	9.20	16.58
Hawaii													
Honolulu	14.08	16.92	19.42	21.24	22.83	23.51	23.74	23.28	21.35	18.06	14.88	13.40	19.42
Idaho													
Boise	5.79	8.97	13.63	18.97	23.51	26.01	27.37	23.62	18.40	12.26	6.70	5.11	15.90
Illinois													
Chicago	6.47	9.31	12.49	16.47	20.44	22.60	22.03	19.31	15.10	10.79	6.47	5.22	13.85
Rockford	6.70	9.77	12.72	16.58	20.33	22.49	22.15	19.42	15.22	10.79	6.59	5.34	14.08
Springfield	7.50	10.33	13.40	17.83	21.46	23.51	23.05	20.56	16.58	12.26	7.72	6.13	15.10
Indiana													
Indianapolis	7.04	9.99	13.17	17.49	21.24	23.28	22.60	20.33	16.35	11.92	7.38	5.79	14.76
Iowa													
Mason City	6.70	9.77	13.29	16.92	20.78	22.83	22.71	19.76	15.33	10.90	6.59	5.45	14.31
Waterloo	6.81	9.77	13.06	16.92	20.56	22.83	22.60	19.76	15.33	10.90	6.70	5.45	14.20

Worldwide, Horizontal Surface Solar Radiation Data, [MJ/m²-day]

Position	Jan.	Feb.	Mar.	Apr.	May	June	July	Aug.	Sep.	Oct.	Nov.	Dec.	Average
Kansas													
Dodge City	9.65	12.83	16.69	21.01	23.28	25.78	25.67	22.60	18.40	14.42	10.11	8.40	17.49
Goodland	8.97	11.92	16.13	20.44	22.71	25.78	25.55	22.60	18.28	14.08	9.65	7.84	17.03
Kentucky													
Lexington	7.27	9.88	13.51	17.60	20.56	22.26	21.46	19.65	16.01	12.38	7.95	6.25	14.54
Louisville	7.27	10.22	13.63	17.83	20.90	22.71	22.03	20.10	16.35	12.38	7.95	6.25	14.76
Louisiana													
New Orleans	9.77	12.83	16.01	19.87	21.80	22.03	20.67	19.65	17.60	15.56	11.24	9.31	16.35
Lake Charles	9.77	12.83	16.13	19.31	21.58	22.71	21.58	20.33	18.06	15.56	11.47	9.31	16.58
Maine													
Portland	6.70	9.99	13.78	16.92	19.99	21.92	21.69	19.31	15.22	10.56	6.47	5.45	13.97
Maryland													
Baltimore	7.38	10.33	13.97	17.60	20.21	22.15	21.69	19.19	15.79	11.92	8.06	6.36	14.54
Massachusetts													
Boston	6.70	9.65	13.40	16.92	20.21	22.03	21.80	19.31	15.33	10.79	6.81	5.45	14.08
Michigan													
Detroit	5.91	8.86	12.38	16.47	20.33	22.37	21.92	18.97	14.76	10.11	6.13	4.66	13.63
Lansing	5.91	8.86	12.49	16.58	20.21	22.26	21.92	18.85	14.54	9.77	5.91	4.66	13.51
Minnesota													
Duluth	5.68	9.31	13.74	17.38	20.10	21.46	21.80	18.28	13.29	8.86	5.34	4.43	13.29
Minneapolis	6.36	9.77	13.51	16.92	20.56	22.49	22.83	19.42	14.65	9.99	6.13	4.88	13.97
Rochester	6.36	9.65	13.17	16.58	20.10	22.15	22.15	19.08	14.54	10.11	6.25	5.11	13.74
Mississippi													
Jackson	9.43	12.38	16.13	19.87	22.15	23.05	22.15	19.08	14.54	10.11	6.25	5.11	13.74
Missouri													
Columbia	8.06	10.90	14.31	18.62	21.58	23.62	23.85	21.12	16.69	12.72	8.29	6.70	15.56
Kansas City	7.95	10.68	14.08	18.28	21.24	23.28	23.62	20.78	16.58	12.72	8.40	6.70	15.44
Springfield	8.52	11.02	14.65	18.62	21.24	23.05	23.62	21.24	16.81	13.17	8.86	7.27	15.67
St. Louis	7.84	10.56	13.97	18.06	21.12	23.05	22.94	20.44	16.58	12.49	8.18	6.59	15.22

Worldwide, Horizontal Surface Solar Radiation Data, [MJ/m²-day]

Position	Jan.	Feb.	Mar.	Apr.	May	June	July	Aug.	Sep.	Oct.	Nov.	Dec.	Average
Montana													
Helena	5.22	8.29	12.61	17.15	20.67	23.28	25.21	21.24	15.79	10.45	6.02	4.43	14.20
Lewistown	5.22	8.40	12.72	17.15	20.33	23.05	24.53	20.78	15.10	10.22	5.91	4.32	13.97
Nebraska													
Omaha	7.50	10.33	13.97	18.06	21.24	2.40	23.51	20.56	16.01	11.81	7.61	6.13	15.10
Lincoln	7.33	10.10	13.65	16.22	19.26	21.21	22.15	18.87	15.44	11.54	7.76	6.20	14.16
Nevada													
Elko	7.61	10.56	14.42	18.85	22.71	25.67	26.69	23.62	19.31	13.63	8.29	6.70	16.58
Las Vegas	10.79	14.42	19.42	24.87	28.16	30.09	28.28	25.89	22.15	17.03	12.15	9.88	20.33
Reno	8.29	11.58	16.24	21.24	25.10	27.48	28.16	24.98	20.56	14.88	9.31	7.38	17.94
New Hampshire													
Concord	6.81	10.11	13.97	16.92	20.21	21.80	21.80	19.08	14.99	10.45	6.47	5.45	14.08
New Jersey													
Atlantic City	7.38	10.22	13.97	17.49	20.21	21.92	21.24	19.19	15.79	11.92	8.06	6.36	14.54
Newark	6.93	9.77	13.51	17.26	19.76	21.35	21.01	18.85	15.33	11.36	7.27	5.68	13.97
New Mexico													
Albuquerque	11.47	14.99	19.31	24.53	27.60	29.07	27.03	24.76	21.12	17.03	12.49	10.33	19.99
New York													
Albany	6.36	9.43	12.95	16.69	19.53	21.46	21.58	18.51	14.65	10.11	6.13	5.00	13.51
Buffalo	5.68	8.40	12.15	16.35	19.76	22.03	21.69	18.62	14.08	9.54	5.68	4.54	13.29
New York City	6.93	9.88	13.85	17.72	20.44	22.03	21.69	19.42	15.56	11.47	7.27	5.79	14.31
Rochester	5.68	8.52	12.26	16.58	19.87	21.92	21.69	18.51	14.20	9.54	5.68	4.54	13.29
North Carolina													
Charlotte	8.97	11.81	15.67	19.76	21.58	22.60	21.92	19.99	16.92	13.97	9.99	8.06	16.01
Wilmington	9.31	12.15	16.24	20.44	21.92	22.60	21.58	19.53	16.69	14.08	10.56	8.52	16.13
North Dakota													
Fargo	5.79	9.09	13.17	16.92	20.56	22.37	23.17	19.87	14.31	9.54	5.68	4.54	13.74
Bismarck	6.12	9.75	13.88	17.43	21.45	23.01	24.06	20.12	15.21	10.61	6.28	4.84	14.39

Worldwide, Horizontal Surface Solar Radiation Data, [MJ/m²-day]

Position	Jan.	Feb.	Mar.	Apr.	May	June	July	Aug.	Sep.	Oct.	Nov.	Dec.	Average
Ohio													
Cleveland	5.79	8.63	12.04	16.58	20.10	22.15	21.92	18.97	14.76	10.22	6.02	4.66	13.51
Columbus	6.47	9.09	12.49	16.58	19.76	21.58	21.12	18.97	15.44	11.24	6.81	5.34	13.74
Dayton	6.81	9.43	12.83	17.03	20.33	22.37	22.37	19.65	15.90	11.47	7.04	5.45	14.20
Youngstown	5.79	8.40	11.92	15.90	19.19	21.24	20.78	18.06	14.31	10.11	6.02	4.77	13.06
Oklahoma													
Oklahoma City	9.88	1.25	16.47	20.33	22.26	24.42	24.98	22.49	18.17	14.54	10.45	8.74	17.15
Oregon													
Eugene	4.54	7.04	11.24	15.79	19.99	22.37	24.19	21.01	15.90	9.65	5.11	3.75	13.40
Medford	5.34	8.52	13.17	18.62	23.39	26.23	27.82	23.96	18.62	11.92	6.02	4.43	15.67
Portland	4.20	6.70	10.68	15.10	18.97	21.24	22.60	19.53	14.88	9.20	4.88	3.52	12.61
Pacific Islands													
Guam	16.35	17.38	19.65	20.78	20.56	19.76	18.28	17.49	17.49	16.58	15.79	15.10	17.94
Pennsylvania													
Philadelphia	7.04	9.88	13.63	17.26	19.99	22.03	21.46	19.42	15.67	11.58	7.72	6.02	14.31
Pittsburgh	6.25	8.97	12.61	16.47	19.65	21.80	21.35	18.85	15.10	10.90	6.59	5.00	13.63
Rhode Island													
Providence	6.70	9.65	13.40	16.92	19.99	21.58	21.24	18.85	15.22	11.02	6.93	5.56	13.97
South Carolina													
Charleston	9.77	12.72	16.81	21.12	22.37	22.37	21.92	19.65	16.92	14.54	11.02	9.09	16.58
Greenville	9.20	12.04	15.90	19.99	21.58	22.60	21.58	19.87	16.81	14.08	10.22	8.18	16.01
South Dakota													
Pierre	6.47	9.54	13.85	17.94	21.46	24.08	24.42	21.46	16.35	11.24	7.04	5.45	14.99
Rapid City	6.70	9.88	14.20	18.28	21.46	24.19	24.42	21.80	16.92	11.81	7.50	5.79	15.33
Tennessee													
Memphis	8.86	11.58	15.22	19.42	22.03	23.85	23.39	21.46	17.38	14.20	9.65	7.84	16.24
Nashville	8.29	11.13	14.65	19.31	21.69	23.51	22.49	20.56	16.81	13.51	8.97	7.15	15.67

Worldwide, Horizontal Surface Solar Radiation Data, [MJ/m²-day]

Position	Jan.	Feb.	Mar.	Apr.	May	June	July	Aug.	Sep.	Oct.	Nov.	Dec.	Average
Texas													
Austin	10.68	13.63	17.03	19.53	21.24	23.74	24.42	22.83	18.85	15.67	11.92	9.99	17.49
Brownsville	10.33	13.17	16.47	19.08	20.78	22.83	23.28	21.58	18.62	16.13	12.38	9.88	17.03
El Paso	12.38	16.24	20.90	25.44	28.05	28.85	26.46	24.30	21.12	17.72	13.63	11.47	20.56
Houston	9.54	12.26	15.22	18.06	20.21	21.69	21.35	20.21	17.49	15.10	11.02	8.97	15.90
San Antonio	10.88	13.53	16.26	17.35	21.10	23.87	24.92	22.81	19.22	15.52	11.50	9.98	17.24
Utah													
Salt Lake City	6.93	10.45	14.76	19.42	23.39	26.46	26.35	23.39	18.85	13.29	8.06	6.02	16.47
Vermont													
Burlington	5.79	9.20	13.06	16.47	19.87	21.69	21.80	18.74	14.42	9.43	5.56	4.43	13.40
Virginia													
Norfolk	8.06	10.90	14.65	18.51	20.78	22.15	21.12	19.42	16.13	12.49	9.09	7.27	15.10
Richmond	8.06	10.90	14.76	18.62	20.90	22.49	21.58	19.53	16.24	12.61	8.97	7.15	15.22
Washington													
Olympia	3.63	6.02	9.99	14.20	18.06	20.10	21.12	18.17	13.63	7.95	4.32	3.07	11.70
Seattle	3.52	5.91	10.11	14.65	19.08	20.78	21.80	18.51	13.51	7.95	4.20	2.84	11.92
Yakima	4.88	7.95	12.83	17.83	22.49	24.87	25.89	22.26	16.92	10.68	5.56	4.09	17.76
West Virginia													
Charleston	7.04	9.65	13.40	17.15	20.21	21.69	20.90	18.97	15.56	11.81	7.72	6.02	14.20
Elkins	6.93	9.43	12.83	16.35	19.08	20.56	19.99	18.06	14.88	11.13	7.27	5.79	13.51
Wisconsin													
Green Bay	6.25	9.31	13.17	16.81	20.56	22.49	22.03	18.85	14.20	9.65	5.79	4.88	13.74
Madison	6.59	9.88	13.29	16.92	20.67	22.83	22.37	19.42	14.76	3.41	6.25	5.22	14.08
Milwaukee	6.47	9.31	12.72	16.69	20.78	22.94	22.60	19.42	14.88	10.22	6.25	5.11	13.97
Wyoming													
Rock Springs	7.61	10.90	15.10	19.42	23.17	26.01	25.78	22.94	18.62	13.40	8.40	6.70	16.58
Sheridan	6.47	9.77	13.97	17.94	20.90	23.85	24.64	21.69	16.47	11.24	7.15	5.56	14.99

(Source: National Renewable Energy Laboratory, USA; Internet Address: http://rredc.nrel.gov/solar)

31

olar Radiation on ilted Surfaces

e worldwide horizontal surface data on pages 18-31 can be ed to estimate solar radiation on tilted surfaces. This section tlines a procedure to use this data to estimate the solar diation on equatorial-facing, tilted collectors. Nomographs are ovided for common tilt angles of: tilt equal to the site latitude d the latitude +/- 15°.

ocedure:

Obtain the monthly averaged, daily total radiation, $\overline{H}_h$, for the chosen location and month of interest, pages 18-31.

Compute the monthly clearness index, $\overline{K}_T$:

$\overline{K}_T = \dfrac{\overline{H}_h}{\overline{H}_{o,h}}$ Where $\overline{H}_{o,h}$ is the extraterrestrial horizontal surface radiation and can be read from Fig. c-1 on page 33.

Determine the sunset hour angle, h_{ss}, from Fig. c-2 page 34, (positive value).

Compute the diffuse to total radiation ratio, $\dfrac{\overline{D}_h}{\overline{H}_h}$:

$$\frac{\overline{D}_h}{\overline{H}_h} = 0.775 + 0.347\left(h_{ss} - \frac{\pi}{2}\right) - \left[0.505 + 0.0261\left(h_{ss} - \frac{\pi}{2}\right)\right]\cos\left(2\overline{K}_T - 1.8\right)$$

Note: h_{ss} in radians.

Solve for the diffuse and beam radiation components:

$$\overline{D}_h = \left(\frac{\overline{D}_h}{\overline{H}_h}\right)\overline{H}_h \quad \text{and} \quad \overline{B}_h = \overline{H}_h - \overline{D}_h$$

6. **Determine the collector tilt factor, $\overline{R}_b$, for the tilt angle β, of interest. Use the following formulation or Figs. c-3-5 for common tilt orientations.**

$$\overline{R}_b = \frac{\cos(L-\beta)\cos(\delta_s)\sin(h_{sr}) + h_{sr}\sin(L-\beta)\sin(\delta_s)}{\cos(L)\cos(\delta_s)\sin(h_{sr}) + h_{sr}\sin(L)\sin(\delta_s)}$$

Where δ_s is the declination angle, page 4, and h_{sr} is the hour angle at sunrise, Fig. c-2 (negative value, radians).

7. **Compute the collector monthly-averaged radiation total, $\overline{H}_c$:**

$$\overline{H}_c = \overline{R}_b\overline{B}_h + \overline{D}_h\cos^2\left(\frac{\beta}{2}\right) + (\overline{D}_h + \overline{B}_h)\rho\sin^2\left(\frac{\beta}{2}\right)$$

Assuming an appropriate reflectivity value, ρ, from the table on page 36.

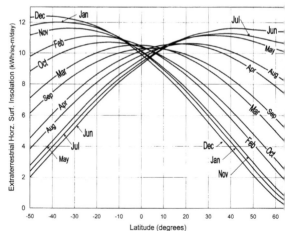

Figure c-1. $\overline{H}_{o,h}$, extraterrestrial ,monthly averaged, daily insolation on a horizontal surface.

33

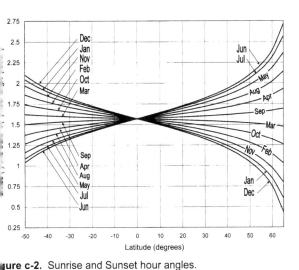

Figure c-2. Sunrise and Sunset hour angles.

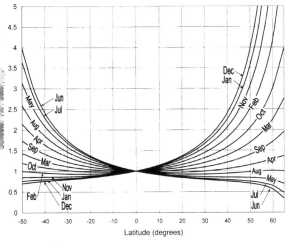

Figure c-3. $\overline{R}_b$ for tilt = L.

34

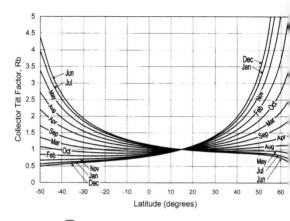

Figure c-4. $\overline{R}_b$ for tilt = L − 15.

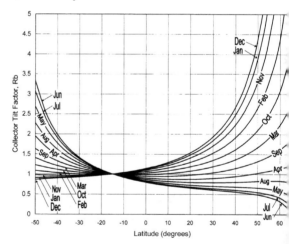

Figure c-5. $\overline{R}_b$ for tilt = L + 15.

Reflectivity values for characteristic surfaces (integrated over solar spectrum and angle of incidence)[a].

Surface	Average reflectivity
Snow (freshly fallen or with ice film)	0.75
Water surfaces (relatively large incidence angles)	0.07
Soils (clay, loam, etc.)	0.14
Earth roads	0.04
Coniferous forest (winter)	0.07
Forests in autumn, ripe field crops, plants	0.26
Weathered blacktop	0.10
Weathered concrete	0.22
Dead leaves	0.30
Dry grass	0.20
Green grass	0.26
Bituminous and gravel roof	0.13
Crushed rock surface	0.20
Building surfaces, dark (red brick, dark paints, etc.)	0.27
Building surfaces, light (light brick, light paints, etc.)	0.60

[a]From Hunn, B. D., and D. O. Calafell, Determination of Average Ground Reflectivity for Solar Collectors, *Sol. Energy,* vol. 19, p. 87, 1977; see also R. J. List, "Smithsonian Meteorological Tables," 6th ed., Smithsonian Institution Press, pp. 442–443, 1949.

Estimation of UV Insolation

Nomenclature

global horizontal total solar radiation
beam component of total solar radiation
beam component of UV radiation
global horizontal UV radiation
cloudiness index
air mass
solar altitude angle

$$\frac{V,h}{t} = 0.14315K_t^2 - 0.20445K_t + 0.135544$$

$$\frac{V,b}{t,b} = 0.0688e^{-0.575m} = 0.688\exp\left(\frac{-0.575}{\sin\alpha}\right)$$

Radiation Heat Transfer Properties

Surface emissivity, ε, which is the emission of the surface compared to an ideal blackbody, E/E_b.

$$\varepsilon \equiv \frac{E}{E_b} = \frac{1}{\sigma T^4} \int_0^\infty \varepsilon_\lambda E_{b\lambda} d\lambda$$

Where σ is the Stefan-Boltzmann constant, (5.670×10^{-8} W/m K^4), T is surface temperature and the subscript λ refers to wavelength.

Incident radiation balance: $\alpha + \tau + \rho = 1$

Where:

absorptivity: $\alpha \equiv \dfrac{I_{absorbed}}{I}$ transmissivity: $\tau \equiv \dfrac{I_{trans.}}{I}$

reflectivity: $\rho \equiv \dfrac{I_{refl.}}{I}$ and I is the total surface irradiation.

Emissivity and absorptivity of common materials

Substance	Short-wave absorptance	Long-wave emittance	α ε
Class I substances: Absorptance to emittance ratios less than 0.5			
Magnesium carbonate, MgCO$_3$	0.025–0.04	0.79	0.03–0
White plaster	0.07	0.91	0.08
Snow, fine particles, fresh	0.13	0.82	0.16
White paint, 0.017 in, on aluminum	0.20	0.91	0.22
Whitewash on galvanized iron	0.22	0.90	0.24
White paper	0.25–0.28	0.95	0.26–0
White enamel on iron	0.25–0.45	0.9	0.28–0.
Ice, with sparse snow cover	0.31	0.96–0.97	0.32
Snow, ice granules	0.33	0.89	0.37
Aluminum oil base paint	0.45	0.90	0.50
White powdered sand	0.45	0.84	0.54

ustance	Short-wave absorptance	Long-wave emittance	α ε
ss II substances: Absorptance to emittance ratios between 0.5 and 0.9			
estos felt	0.25	0.50	0.50
en oil base paint	0.5	0.9	0.56
cks, red	0.55	0.92	0.60
estos cement board, white	0.59	0.96	0.61
rble, polished	0.5–0.6	0.9	0.61
od, planed oak	—	0.9	—
ugh concrete	0.60	0.97	0.62
ncrete	0.60	0.88	0.68
ass, green, after rain	0.67	0.98	0.68
ass, high and dry	0.67–0.69	0.9	0.76
getable fields and shrubs, wilted	0.70	0.9	0.78
k leaves	0.71–0.78	0.91–0.95	0.78–0.82
zen soil	—	0.93–0.94	—
sert surface	0.75	0.9	0.83
mmon vegetable fields and shrubs	0.72–0.76	0.9	0.82
ound, dry plowed	0.75–0.80	0.9	0.83–0.89
k woodland	0.82	0.9	0.91
e forest	0.86	0.9	0.96
rth surface as a whole (land and sea, no clouds)	0.83	—	—
ss III substances: Absorptance to emittance ratios between 0.8 and 1.0			
ey paint	0.75	0.95	0.79
d oil base paint	0.74	0.90	0.82
estos, slate	0.81	0.96	0.84
estos, paper		0.93–0.96	—
oleum, red-brown	0.84	0.92	0.91
r sand	0.82	0.90	0.91
en roll roofing	0.88	0.91–0.97	0.93
te, dark grey	0.89	—	—
d grey rubber	—	0.86	—
rd black rubber	—	0.90–0.95	—
phalt pavement	0.93	—	—
ck cupric oxide on copper	0.91	0.96	0.95
e moist ground	0.9	0.95	0.95
t sand	0.91	0.95	0.96
ter	0.94	0.95–0.96	0.98
ck tar paper	0.93	0.93	1.0
ck gloss paint	0.90	0.90	1.0
all hole in large box, furnace, or enclosure	0.99	0.99	1.0
ohlraum," theoretically perfect black body	1.0	1.0	1.0

Emissivity and absorptivity of common materials,
continued from page 37.

Substance	Short-wave absorptance	Long-wave emittance	α ϵ
Class IV substances: Absorptance to emitance ratios greater than 1.0			
Black silk velvet	0.99	0.97	1.02
Alfalfa, dark green	0.97	0.95	1.02
Lampblack	0.98	0.95	1.03
Black paint, 0.017 in. on aluminum	0.94–0.98	0.88	1.07–1
Granite	0.55	0.44	1.25
Graphite	0.78	0.41	1.90
High ratios, but absorptances less than 0.80			
Dull brass, copper, lead	0.2–0.4	0.4–0.65	1.63–2
Galvanized sheet iron, oxidized	0.8	0.28	2.86
Galvanized iron, clean, new	0.65	0.13	5.0
Aluminum foil	0.15	0.05	3.00
Magnesium	0.3	0.07	4.3
Chromium	0.49	0.08	6.13
Polished zinc	0.46	0.02	23.0
Deposited silver (optical reflector) untarnished	0.07	0.01	
Class V substances: Selective surfaces[b]			
Plated metals:[c]			
Black sulfide on metal	0.92	0.10	9.2
Black cupric oxide on sheet aluminum	0.08–0.93	0.09–0.21	
Copper (5×10^{-5} cm thick) on nickel or silver-plated metal			
Cobalt oxide on platinum			
Cobalt oxide on polished nickel	0.93–0.94	0.24–0.40	3.9
Black nickel oxide on aluminum	0.85–0.93	0.06–0.1	14.5–15
Black chrome	0.87	0.09	9.8
Particulate coatings:			
Lampblack on metal			
Black iron oxide, 47 μm grain size, on aluminum			
Geometrically enhanced surfaces:[d]			
Optimally corrugated greys	0.89	0.77	1.2
Optimally corrugated selectives	0.95	0.16	5.9
Stainless-steel wire mesh	0.63–0.86	0.23–0.28	2.7–3.0
Copper, treated with $NaClO_2$ and NaOH	0.87	0.13	6.69

[a]From Anderson, B., "Solar Energy," McGraw-Hill Book Company, 1977, with permission.

[b]Selective surfaces absorb most of the solar radiation between 0.3 and 1.9 μm, and emit very little the 5–15 μm range—the infrared.

[c]For a discussion of plated selective surfaces, see Daniels, "Direct Use of the Sun's Energy," especia chapter 12.

[d]For a discussion of how surface selectivity can be enhanced through surface geometry, see K. G. Hollands, Directional Selectivity Emittance and Absorptance Properties of Vee Corrugated Specular S faces, *J. Sol. Energy Sci. Eng.*, vol. 3, July 1963.

Properties of some selective plated coating systems[a]

Coating[b]	Substrate	$\bar{\alpha}_s$	$\bar{\epsilon}_i$	Durability Breakdown temperature (°C)	Durability Humidity-Degradation MIL STD 810B
Black nickel on nickel	Steel	0.95	0.07	>290	Variable
Black chrome on nickel	Steel	0.95	0.09	>430	No effect
Black chrome	Steel	0.91	0.07	>430	Completely rusted
	Copper	0.95	0.14	315	Little effect
	Galvanized steel	0.95	0.16	>430	Complete removal
Black copper	Copper	0.88	0.15	315	Complete removal
Iron oxide	Steel	0.85	0.08	430	Little effect
Manganese oxide	Aluminum	0.70	0.08		
Organic overcoat on iron oxide	Steel	0.90	0.16		Little effect
Organic overcoat on black chrome	Steel	0.94	0.20		Little effect

[a]From U.S. Dept. of Commerce, "Optical Coatings for Flat Plate Solar Collectors," NTIS No. PN-252-383, Honeywell, Inc., 1975.
[b]Black nickel coating plated over a nickel-steel substrate has the best selective properties ($\bar{\alpha}_s$ = 0.95, $\bar{\epsilon}_i$ = 0.07), but these degraded significantly during humidity tests. Black chrome plated on a nickel-steel substrate also had very good selective properties ($\bar{\alpha}_s$ = 0.95, $\bar{\epsilon}_i$ = 0.09) and also showed high resistance to humidity.

.40

Angular variation of the absorptivity of lampblack paint

Incidence angle $i(°)$	Absorptance $\alpha(i)$
0–30	0.96
30–40	0.95
40–50	0.93
50–60	0.91
60–70	0.88
70–80	0.81
80–90	0.66

Adapted from [9]

Reflectivity values for reflector materials

Material	ρ
Silver (unstable as front surface mirror)	0.94 ± 0.02
Gold	0.76 ± 0.03
Aluminized acrylic, second surface	0.86
Anodized aluminum	0.82 ± 0.05
Various aluminum surfaces-range	0.82–0.92
Copper	0.75
Back-silvered water-white plate glass	0.88
Aluminized type-C Mylar (from Mylar side)	0.76

Spectral Absorption of Solar Radiation in Water

Wavelength (μm)	Layer depth				
	0	1 cm	10 cm	1 m	10 m
0.2–0.6	23.7	23.7	23.6	22.9	17.
0.6–0.9	36.0	35.3	36.0	12.9	0.
0.9–1.2	17.9	12.3	0.8	0.0	0.
1.2 and over	22.4	1.7	0.0	0.0	0.
Total	100.0	73.0	54.9	35.8	18.

*Numbers in the table give the percentage of sunlight in the wavelength band passing through water to the indicated thickness.

41

ansparent Materials

fraction of light between materials

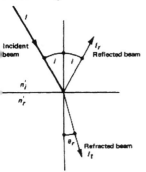

ex of refraction:

$$\frac{\sin(i)}{n(\theta_r)} = \frac{n'_r}{n'_i} = n_r$$

sible spectrum refractive index values, n_r, based on air

aterial	Index of Refraction
r	1.000
ean polycarbonate	1.59
amond	2.42
ass (solar collector type)	1.50–1.52
exiglass[a] (polymethyl methacrylate,	
MA)	1.49
ylar[a] (polyethylene terephthalate, PET)	1.64
artz	1.54
dlar[a] (polyvinyl fluoride, PVF)	1.45
flon[a] (polyfluoroethylenepropylene, FEP)	1.34
ter–liquid	1.33
ter–solid	1.31

[a]Trademark of the duPont Company, Wilmington, Delaware.

Thermal and radiative properties of collector cover materials[b]

Material name	Index of refraction (n)	τ (solar)[b] (%)	τ (solar) (%)	τ (infrared)[b] (%)	Expansion coefficient (m/m × °C)	Temperature limits (°C)	Weatherability (comment)	Chemical resistance (comment)
Lexan (polycarbonate)	1.586 (D 542)	125 mil 64.1 (±0.8)	125 mil 72.6 (±0.1)	125 mil 2.0	6.75 (10^{-5}) (H 696)	121–132 service temperature	Good: 2 yr exposure in Florida caused yellowing; 5 yr caused 5% loss in τ	Good: comparable to acrylic
Plexiglas (acrylic)	1.49 (D 542)	125 mil 89.6 (±0.3)	125 mil 79.6 (±0.8)	(est)[y] 2.0	7.02 (10^{-5}) at 15.5°C; 8.28 (10^{-5}) at 38°C	82–93 service temperature	Average to good: based on 20 yr testing in Arizona, Florida, and Pennsylvania	Good to excellent: resists most acids and alkalis
Teflon F.E.P. (fluorocarbon)	1.343 (D 542)	5 mil 92.3 (±0.2)	5 mil 89.8 (±0.4)	5 mil 25.6 (±0.5)	1.06 (10^{-4}) at 71°C; 1.6 (10^{-4}) at 100°C	204 continuous use; 246 short-term use	Good to excellent: based on 15 yr exposure in Florida environment	Excellent: chemically inert
Tedlar P.V.F. (fluorocarbon)	1.46 (D 542)	4 mil 92.2 (±0.1)	4 mil 88.3 (±0.9)	4 mil 20.7 (±0.2)	5.04 (10^{-5}) (D 696)	107 continuous use; 177 short-term use	Good to excellent: 10 yr exposure in Florida with slight yellowing	Excellent: chemically inert
Mylar (polyester)	1.64–1.67 (D 542)	5 mil 86.9 (±0.3)	5 mil 80.1 (±0.1)	5 mil 17.8 (±0.5)	1.69 (10^{-5}) (D 696-44)	150 continuous use; 204 short-term use	Poor: ultraviolet degradation great	Good to excellent: comparable to Tedlar
Sunlite (fiberglass)	1.54 (D 542)	25 mil (P) 86.5 (±0.2) 25 mil (R) 87.5 (±0.2)	25 mil (P) 75.4 (±0.1) 25 mil (R) 77.1 (±0.7)	25 mil (P) 7.6 (±0.1) 25 mil (R) 3.3 (±0.3)	2.5 (10^{-5}) (D 696)	93 continuous use; causes 5% loss in τ	Fair to good: regular, 7 yr solar life; premium, 20 yr solar life	Good: inert to chemical atmospheres
Float glass (glass)	1.518 (D 542)	125 mil 84.3 (±0.1)	125 mil 78.5 (±0.2)	125 mil 2.0	8.64 (10^{-5}) (D 696)	732 softening point; 38 thermal shock	Excellent: time proved	Good to excellent: time proved

Thermal and radiative properties of collector cover materials[a]

Material name	Index of refraction (n)	τ (solar)[b] (%)	τ (solar)[c] (%)	τ (infrared)[b] (%)	Expansion coefficient (m/m × °C)	Temperature limits (°C)	Weatherability (comment)	Chemical resistance (comment)
Temper glass (glass)	1.518 (D 542)	125 mil 84.3 (±0.11)	125 mil 78.6 (±0.2)	125 mil 2.0 (est)[d]	8.64 (10-6) (D696)	232-260 continuous use; 260-288 short-term use	Excellent: time proved	Good to excellent: time proved
Clear lime sheet glass (low iron oxide glass)	1.51 (D 542)	Insufficient data provided by ASG	125 mil 87.5 (±0.5)	125 mil 2.0 (est)	9 (10-6) (D 696)	204 for continuous operation	Excellent: time proved	Good to excellent: time proved
Clear lime temper glass (low iron oxide glass)	1.51 (D 542)	Insufficient data provided by ASG	125 mil 87.5 (±0.5)	125 mil 2.0 (est)	9 (10-6) (D 696)	204 for continuous operation	Excellent: time proved	Good to excellent: time proved
Sunadex white crystal glass (0.01% iron oxide glass)	1.50 (D 542)	Insufficient data provided by ASG	125 mil 91.5 (±0.2)	125 mil 2.0 (est)	8.46 (10-6) (D 696)	204 for continuous operation	Excellent: time proved	Good to excellent: time proved

[a]Numerical integration $(\Sigma \tau_{\lambda \text{rep}} F_{\lambda, T \to \lambda_1 T})$ for λ = 0.2–4.0 μM.

[b]Numerical integration $(\Sigma \tau_{\lambda \text{rep}} F_{\lambda, T \to \lambda_o})$ for λ = 3.0–50.0 μM.

[c]All parenthesized numbers refer to ASTM test codes.

[d]Data not provided; estimate of 2% to be used for 125 mil samples.

[e]Degrees differential to rupture 2 × 2 × ¼ in samples. Glass specimens heated and then quenched in water bath at 70°F.

[f]Sunlite premium data denoted by (P); Sunlite regular data denoted by (R).

[g]Compiled data based on ASTM Code E 424 Method B.

[h]Abstracted from Ratzel, A. C., and R. B. Bannerot, Optimal Material Selection for Flat-Plate Solar Energy Collectors Utilizing Commercially Available Materials, presented at ASME-AIChE Natl. Heat Transfer Conf., 1976.

Solar Thermal Collector Overview

General Configuration	Description	Concentration Ratio	Indicative Operating Temp. (°C)
	Non-Convecting Solar Pond	1	30-70
	Unglazed Flat Plate Absorber	1	O 40
	Flat Plate Collector (High Efficiency)	1 (1)	O 70 (60-120)
	Fixed Concentrator	3-5	100-150
	Evacuated Tube	1	50-180
	Compound Parabolic (With 1 Axis Tracking)	1-5 (5-15)	70-240 (70-290)
	Parabolic Trough	10-50	150-350
	Fresnel Refractor	10-40	70-270
	Spherical Dish Reflector	100-300	70-730
	Parabolic Dish Reflector	200-500	250-700
	Central Receiver	500-3000	500->100

Adapted from

Non-Concentrating Collectors

iciency of N-C collectors, Hottel-Whillier equation (European andard on page 48):

$$\eta_c = F_R \tau_s \alpha_s - F_R U_c \frac{\left(T_{f,in} - T_a\right)}{I_c}$$

here F_R is the collector heat removal factor, τ_s is the cover nsmissivity, α_s is the cover-absorber absorptivity, and U_c is the erall collector heat loss conductance. $T_{f,in}$ is the collector fluid et temperature, T_a is the ambient, and I_c is the incident diation on the collector.

presentative performance curves

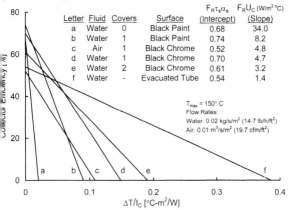

Letter	Fluid	Covers	Surface	$F_R\tau_s\alpha_s$ (Intercept)	$F_R U_c$ (W/m²°C) (Slope)
a	Water	0	Black Paint	0.68	34.0
b	Water	1	Black Paint	0.74	8.2
c	Air	1	Black Chrome	0.52	4.8
d	Water	1	Black Chrome	0.70	4.7
e	Water	2	Black Chrome	0.61	3.2
f	Water	-	Evacuated Tube	0.54	1.4

T_{max} = 150° C
Flow Rates:
Water: 0.02 kg/s/m² (14.7 lb/h/ft²)
Air: 0.01 m³/s/m² (19.7 cfm/ft²)

cidence Angle Modifier

cidence angle modifier, $K_{\tau\alpha}$, is used to estimate collector rformance at non-normal angles of incidence with:

$$\eta_c = F_R \left[K_{\tau\alpha} \left(\tau_s \alpha_s\right)_n - U_c \frac{\left(T_{f,in} - T_a\right)}{I_c} \right]$$

46

where $K_{\tau\alpha}$ is of the form (b = constant):

$$K_{\tau\alpha} = 1 - b\left(\frac{1}{\cos(i)} - 1\right)$$

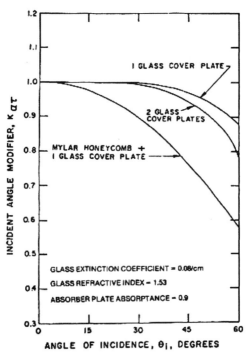

Incident angle modifier for three flat-plate solar collectors.
Reprinted by permission of the American Society of Heating,
Refrigerating and Air-Conditioning Engineers, Inc., Atlanta, from
ASHRAE Standard 93-77, "Methods of Testing to Determine the
Thermal Performance of Solar Collectors."

ropean Standard N-C collector efficiency formulation:

$$= F_R \tau_s \alpha_s - a_1 \frac{T_{ave} - T_{amb}}{G} - a_2 \frac{\left(T_{ave} - T_{amb}\right)^2}{G}$$

here $T_{ave} = 0.5(T_{f,in} + T_{f,out})$, is used instead of collector inlet nperature and G is the hemispherical irradiance (W/m^2).

e incidence angle modifier is applied in a similar manner:

$$\alpha_s = K_{\tau\alpha} \left(\tau_s \alpha_s \right)_n$$

oncentrating Thermal Collectors

ometric Concentration Ratio, CR:

$$R = \frac{A_a}{A_r} \qquad \text{where } A_a \text{ is the aperture area and } A_r \text{ is the}$$

:eiver area.

stantaneous collector efficiency, η_c :

$$= \eta_o - U_c \frac{\left(T_r - T_a\right)}{I_c \left(CR\right)}$$

here η_o is the optical efficiency, U_c is the overall collector

at loss conductance, T_r is the receiver temperature, T_a is the bient temperature, and I_c is the incident radiation on the lector.

Incidence factors for various orientation and tracking arrangements of concentrating collectors

Orientation of collector	Incidence factor $\cos i$
Fixed, horizontal, plane surface.	$\sin L \sin \delta_s + \cos \delta_s \cos h_s \cos L$
Fixed plane surface tilted so that it is normal to the solar beam at noon on the equinoxes.	$\cos \delta_s \cos h_s$
Rotation of a plane surface about a horizontal east-west axis with a single daily adjustment permitted so that its surface normal coincides with the solar beam at noon every day of the year.	$\sin^2 \delta_s + \cos^2 \delta_s \cos h_s$
Rotation of a plane surface about a horizontal east-west axis with continuous adjustment to obtain maximum energy incidence.	$\sqrt{1 - \cos^2 \delta_s \sin^2 h_s}$
Rotation of a plane surface about a horizontal north-south axis with continuous adjustment to obtain maximum energy incidence.	$[(\sin L \sin \delta_s + \cos L \cos \delta_s \cos h_s + \cos^2 \delta_s \sin^2 h_s]^{1/2}$
Rotation of a plane surface about an axis parallel to the earth's axis with continuous adjustment to obtain maximum energy incidence.	$\cos \delta_s$
Rotation about two perpendicular axes with continuous adjustment to allow the surface normal to coincide with the solar beam at all times.	1

*The incidence factor denotes the cosine of the angle between the surface normal and the solar beam

Adaptation of monthly-averaged, horizontal surface data for tracking concentrators:

$$\overline{H}_c = \left[\overline{r}_T - \overline{r}_d \left(\frac{\overline{D}_h}{\overline{H}_h} \right) \right] \overline{H}_h \quad \text{where:}$$

$\overline{D}_h$ monthly-average diffuse radiation component for a horizontal surface

$\overline{H}_c$ monthly-average terrestrial radiation for a tracking collector

$\overline{H}_h$ monthly-average terrestrial radiation for a horizontal surface

$\overline{r}_d$ diffuse radiation factor

$\overline{r}_T$ tracking factor

Thermal Energy Storage

Sensible heat storage:

$$Q_{sensible} = \rho V c_p \Delta T$$

Combined latent and sensible heat storage:

$$Q_{total} = m\left[\overline{c}_{p_{sol}}\left(T_{melt} - T_{low}\right) + \lambda + \overline{c}_{p_{liquid}}\left(T_{high} - T_{melt}\right)\right]$$

Thermochemical energy storage:

$$Q_{thermochemical} = a_r m \Delta H$$

Where m is the mass of reactant, a_r is the fraction reacted and is the heat of reaction per unit mass.

Storage materials properties given on pages 51-53.

Thermal conductivities of containment materials

Materials	Thermal conductivity[b]	
	(W/m K)	(Btu in/hr ft² °F)
Plastics[a]		
ABS	0.17–0.33	1.2–2.3
Acrylic	0.19–0.43	1.3–3.0
Polypropylene	0.12–0.17	0.8–1.2
Polyethylene (high density)	0.43–0.52	3.0–3.6
Polyethylene (medium density)	0.30–0.42	2.1–2.9
Polyethylene (low density)	0.30	2.1
Polyvinyl chloride	0.13	0.9
Metals[c]		
Aluminum	200	1500
Copper	390	2700
Steel	48	330

[a] *Plastics, a desk-top data bank,* 5th Ed. Book A. San Diego, CA: The International Plastics Selector, , 1980.
[b] As measured by ASTM C-177.
[c] *Handbook of Chemistry and Physics,* 40th Ed.

Physical properties of some sensible heat storage materials

Storage Medium	Temperature Range, °C	Density (ρ), kg/m³	Specific Heat (C), J/kg K	Energy Density (ρC) kWh/m³ K	Thermal Conductivity (W/m K)
Water	0–100	1000	4190	1.16	0.63 at 38°C
Water (10 bar)	0–180	881	4190	1.03	—
50% ethylene glycol-50% water	0–100	1075	3480	0.98	—
Dowtherm A® (Dow Chemical, Co.)	12–260	867	2200	0.53	0.122 at 260°C
Therminol 66® (Monsanto Co.)	–9–343	750	2100	0.44	0.106 at 343°C
Draw salt (50NaNO₃–50KNO₃)ᵃ	220–540	1733	1550	0.75	0.57
Molten salt (53KNO₃/40NaNO₂/7NaNO₃)ᵃ	142–540	1680	1560	0.72	0.61
Liquid Sodium	100–760	750	1260	0.26	67.5
Cast iron	m.p. (1150–1300)	7200	540	1.08	42.0
Taconite	—	3200	800	0.71	—
Aluminum	m.p. 660	2700	920	0.69	200
Fireclay	—	2100–2600	1000	0.65	1.0–1.5
Rock	—	1600	880	0.39	—

ᵃ Composition in percent by weight.
Note: m.p. = melting point.

51

Physical properties of latent heat storage materials or PCMs

Storage Medium	Melting Point °C	Latent Heat, kJ/kg	Specific Heat (kJ/kg °C)		Density (Kg/m³)		Energy Density (kWhr/m³K)	Thermal Conductivity (W/m K)
			Solid	Liquid	Solid	Liquid		
LiClO₃ · 3H₂O	8.1	253	—	—	1720	1530	108	—
Na₂SO₄ · 10H₂O (Glauber's Salt)	32.4	251	1.76	3.32	1460	1330	92.7	2.25
Na₂S₂O₃ · 5H₂O	48	200	1.47	2.39	1730	1665	92.5	0.57
NaCH₃COO · 3H₂O	58	180	1.90	2.50	1450	1280	64	0.5
Ba(OH)₂ · 8H₂O	78	301	0.67	1.26	2070	1937	162	0.653ℓ
Mg(NO₃) · 6H₂O	90	163	1.56	3.68	1636	1550	70	0.611
LiNO₃	252	530	2.02	2.041	2310	1776	261	1.35
LiCO₃/K₂CO₃, (35:65)ᵃ	505	345	1.34	1.76	2265	1960	188	—
LiCO₃/K₂CO₃/Na₂CO₃ (32:35:33)ᵃ	397	277	1.68	1.63	2300	2140	165	0.150
n-Tetradecane	5.5	228	—	—	825	771	48	0.150
n-Octadecane	28	244	2.16	—	814	774	52.5	0.361
HDPE (cross-linked)	126	180	2.88	2.51	960	900	45	
Steric acid	70	203	—	2.35	941	347	48	0.172ℓ

ᵃComposition in percent by weight.
Note: ℓ = liquid.

Properties of thermochemical storage media

Reaction	Condition of Reaction Pressure, kPa	Temperature, °C	Component (Phase)	Pressure, kPa	Temperature, °C	Density, kg/m³	Volumetric Storage Density, kWh/m³
$MgCO_3(s) + 1200$ kJ/kg $=$ $MgO(s) + CO_2(g)$	100	427–327	$MgCO_3(s)$ $CO_2(\ell)$	100 7400	20 31	1500 465	187
$Ca(OH)_2(s) + 1415$ kJ/kg $=$ $CaO(s) + H_2O(g)$	100	572–402	$Ca(OH)_2(s)$ $H_2O(\ell)$	100	20	1115	345
$SO_3(g) + 1235$ kJ/kg $=$ $SO_2(g) + \frac{1}{2}O_2(g)$	100	520–960	$SO_3(\ell)$ $SO_2(\ell)$ $O_2(g)$	100 630 10000	45 40 20	1900 1320 130	280

Note: s = solid; ℓ = liquid; g = gas

53

hotovoltaics
ell Output Characteristics

output power across load, P_L: $P_L = I_L V = I_L^2 R_L$

re I_L is the load circuit current, V is the voltage and R_L is the load resistance.

resentative current, voltage, and power outputs of a PV cell

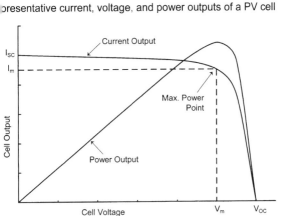

en circuit junction voltage: $V_{oc} = \dfrac{kT}{e_o} \ln\left(\dfrac{I_{sc}}{I_o} + 1\right)$

re:
    k   Boltzman's constant, (1.381 x 10⁻²³ J/K)
    T   cell temperature, (K)
    e_o electron charge, (1.602 x 10⁻¹⁹ J/V)
    I_{sc} short circuit current
    I_o reverse saturation current

tage at maximum power output, V_m, is found from:

$$\exp\left(\frac{e_o V_m}{kT}\right)\left(1 + \frac{e_o V_m}{kT}\right) = 1 + \frac{I_{sc}}{I_o}$$

Effect of illumination and load resistance on PV cell output

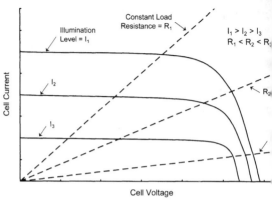

Effect of temperature on cell output

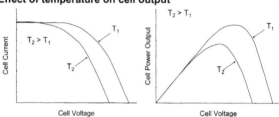

Temperature corrections for the cell output are supplied by the manufacturers and are typically of the form:

$$V = V_{ref}\left(1 + \beta\left(T_c - T_{ref}\right)\right) \quad \text{and} \quad I = I_{ref}\left(1 + \alpha\left(T_c - T_{ref}\right)\right)$$

where the subscript "ref" refers to values at a reference condition and T_c is the actual operating temperature of the cell. α and β are constants provided by the manufacturer. Since α is much smaller than β, power output goes down as the temperature of the cell goes up.

55

Multiple cell output

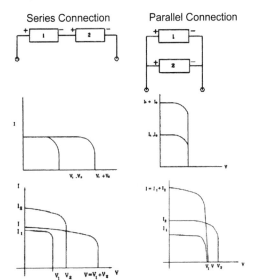

Effect of collector tilt and tracking on incident solar radiation

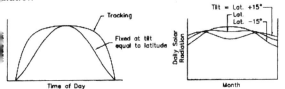

PV Power Configurations

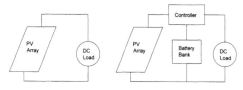

Direct PV-load connection, without and with storage

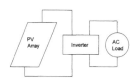

AC power supply

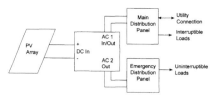

Utility interconnected configuration

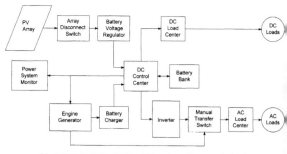

Block diagram for stand-alone PV system with generator backup

Adapted from [10

57

PV System Design

System Design Suggestions from [13]

Keep it simple – Complexity lowers reliability and increases maintenance costs.

Understand system availability – Achieving 99+ % availability with <u>any</u> energy system is expensive.

Be thorough, but realistic, when estimating the load – A large safety factor can cost you a great deal of money.

Cross-check weather sources – Errors in solar resource estimates can cause disappointing system performance.

Know what hardware is available at what cost – Tradeoffs are inevitable. The more you know about hardware, the better decisions you can make. Shop for bargains, talk to dealers, ask questions.

Install the system carefully – Make each connection as if it had to last 30 years—it does. Use the right tools and technique. The system reliability is no higher than its weakest connection.

Safety first and last – Don't take shortcuts that might endanger life or property. Comply with local and national building and electrical codes.

Plan periodic maintenance – PV systems have an enviable record for unattended operation, but no system works forever without some care.

Calculate the life-cycle cost (LCC) to compare PV systems to alternatives – LCC reflects the complete cost of owning and operating any energy system.

Simplified Average Daily Load Determination from [10]

Identify all loads to be connected to the PV system.

For each load, determine its voltage, current, power and daily operating hours. For some loads, the operation may vary on a daily, monthly or seasonal basis. If so, this must be accounted for in calculating daily averages.

Separate ac loads from dc loads.

Determine average daily Ah for each load from current and operating hours data. If operating hours differ from day to day during the week, the daily average over the week should be calculated. If average daily operating hours vary

58

from month to month, then the load calculation may need be determined for each month.

5. Add up the Ah for the dc loads, being sure all are at the same voltage.

6. If some dc loads are at a different voltage, which will require a dc to dc converter, then the converter input Ah these loads needs to account for the conversion efficiency of the converter.

7. For ac loads, the dc input current to the inverter must be determined and the dc Ah are then determined from the input current. The dc input current is determined by equating the ac load power to the dc input power and the dividing by the efficiency of the inverter.

8. Add the Ah for the dc loads to the Ah for the ac loads, the divide by the wire efficiency factor and the battery efficiency factor to obtain the corrected average daily Ah for the total load.

9. The total ac power will determine the required size of the inverter. Individual load powers will be needed to determine wire sizing to the loads. Total load current will be compared with total array current when sizing wire from battery to controller.

Storage Estimation: Estimation of the days of battery storage needed for a stand-alone system if no better estimate is available, [10].

$$D_{crit} = -1.9T_{min} + 18.3 \qquad \text{or}$$

$$D_{non} = -0.48T_{min} + 4.58$$

Where:

D_{crit}	number of storage days required for critical application
D_{non}	number of storage days required for non-critical application
T_{min}	minimum average daily peak sun hours for selected collector tilt during any month of operation

Note: $T_{min} \geq 1$ peak sun hours per day

‍ater Pumping Load Data

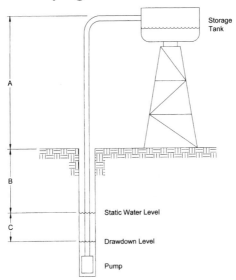

Storage Tank

A

B

Static Water Level

C

Drawdown Level

Pump

mping power (W):

$$P = \frac{\rho_w \dot{V} g H}{\eta_p}$$

here:

gravitational acceleration, (9.81 m/s²)
total pumping head, (m)
dynamic head, (m)
friction head, (m)
static head, (m)

volumetric flow rate, (m³/s)

pump efficiency

water velocity at pipe outlet, (m/s)

water density, (997 kg/m³)

d: $H = H_d + H_s$ $H_d = H_f + \frac{v^2}{2g}$

$$H_s = \begin{cases} A + B, & \text{for no draw down} \\ A + B + C, & \text{for water level drawn down a depth } C \end{cases}$$

Representative water pump operating characteristics

Head (m)	Type pump	Wire-to-water efficiency (%)
0–5	Centrifugal	15–25
6–20	Centrifugal with Jet	10–20
	Submersible	20–30
21–100	Submersible	30–40
	Jack pump	30–45
>100	Jack pump	35–50

Electrical wire load rating

Resistance and amperage ratings for type THHN insulated wire

AWG Wire Size	Resistance @ 20°C (Ω/100 ft or Ω/30.5 m)	Maximum Recommended Current (A)
14	0.2525	15
12	0.1588	20
10	0.9989	30
8	0.6282	55
6	0.3951	75
4	0.2485	95
3	0.1970	110
2	0.1563	130
1	0.1239	150
0	0.0983	170
00	0.0779	195
000	0.0618	225
0000	0.0490	260

Voltage drop due to line resistance:

$$\underbrace{\text{Voltage Drop}}_{\left(\frac{V}{100\,\text{ft}(30.5\,\text{m})}\right)} = \underbrace{\text{Current}}_{(A)} \times \underbrace{\text{Wire Resistance}}_{\left(\frac{\Omega}{100\,\text{ft}(30.5\,\text{m})}\right)}$$

ater Heating Systems

neral design configurations and guidelines. Collector
formance information is available on page 46, system
iluation and sizing can be estimated using the f-chart method,
jinning page 66.

egrated collector and storage or batch systems.
rage tank integrated with collector or storage tank itself is the
ar absorber. Circulation is passive through natural
wection.

  Simple, no moving parts, long lifetime, little maintenance.

  Small systems only, limited freeze protection.

tural circulation (thermosyphon loop) system.

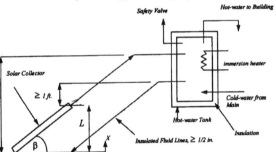

culation is caused by the difference in density, ρ, between the
water in the collector and cooler water exiting the storage
k. To estimate circulation rate, compute the flow pressure
p, ΔP_{flow}:

$$P_{flow} = \rho_{storage} gH - \left[\rho_{collector\ ave} gL + \rho_{coll\ out} g(H-L) \right]$$

w velocity, V, can then be determined by knowing K_{loop}, the
p sum of the component velocity loss factors:

$$V = \sqrt{\frac{\Delta P_{flow}}{\rho_{loop\ ave} K_{loop}}}$$

Natural circulation (thermosyphon loop) system.

👍 Simple, moderate sizes, long lifetime in areas with little chance of freezing.

👎 Tank must be mounted above collectors, freeze protectio difficult.

Forced circulation open loop system.

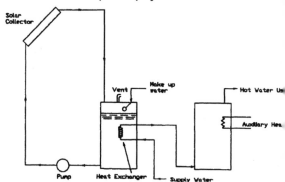

Fluid is actively pumped through the collector, and the reservo is vented so pressure is maintained at atmospheric. Drainbac operation is possible for freeze and stagnation protection.

👍 Simple, increased capacity compared to passive circulati

👎 Freeze protection not as reliable as closed loop drainbac pump must supply entire head from storage tank to collector.

Forced circulation closed loop pressurized system.
Fluid loop is not vented so pumping power is limited to the flo resistance of the piping. Since fluid stays in the collector, glyc solutions are used for freeze protection.

👍 Good freeze protection in cold climates, reduced pumpin power, can be used when drainback is not possible.

👎 Complex system, fluid expansion must be accommodate no stagnation protection.

63

rced circulation closed loop pressurized system.

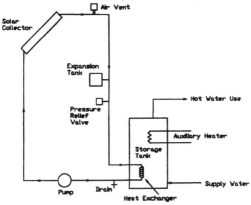

rced circulation closed loop drainback system.

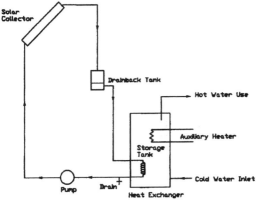

...uid in collectors is allowed to drain into a reservoir located near
...e collectors but in a non-freezing location.

Reliable freeze and stagnation protection, pure water
working fluid can be used, pumping head reduced with
elevated drainback tank.

Piping must have sufficient slope for drainback.

64

Hot Water Load Data

Energy requirement for service water heating:

$$q_{hw} = \rho_w Q c_{pw} \left(T_d - T_s \right)$$

Where:

c_{pw}	specific heat of water, 4.18 kJ/kg-K
q_{hw}	water heating energy requirement
Q	volumetric water flow rate
T_d	water delivery temperature
T_s	water supply temperature
ρ_w	water density, 997 kg/m^3

Guidelines for service hot water demand rates

	Demand per person	
Usage type	liters/day	gal/day
Retail store	2.8	0.75
Elementary school	5.7	1.5
Multifamily residence	76.0	20.0
Single-family residence	76.0	20.0
Office building	11.0	3.0

Building Heat Load Data

Given the overall building loss coefficient, $\overline{UA}$ (W/°C), the degree-day method can be used to estimate the heat demand, Q_n, for day n: $Q_n = \overline{UA}\left(T_{nl} - \overline{T}_a \right)_n$ with: $T_{nl} = T_i - \dfrac{q_i}{\overline{UA}}$

Where T_i is the interior temperature, q_i is the interior heat generation, and $\overline{T}_a$ (daily average temp.) can be estimated from location-specific maximum and minimum temperature data:

$$\overline{T}_a = \frac{T_{a,max} + T_{a,min}}{2}$$

Note: For a thorough treatment of heating loads, the reader is referred to the ASHRAE Handbook, Fundamentals, [1].

eating System Evaluation and Sizing

ocedure for estimating the performance and/or size of
ndard solar heating applications using f-chart. The f-chart
thod computes the solar-supplied fraction, f_s, of thermal
ergy for liquid and air based heating systems.

te: The procedures outlined here are for estimation purposes only, detailed
culations of the system thermal performance and solar collection are needed to
ure satisfactory system performance.

f-chart method assumes standard system configurations, Figs. f-1 and f-2, and
ies only to these systems, with limited variations. For example, the collector-to-
age heat exchanger in the liquid based system (Fig. f-1) may be eliminated (F_{hx} =
r for the air-based system (Fig. f-2) the two-tank domestic water heater may be
nfigured as a one tank system. Furthermore, f-chart is applicable to solar heating
ems where the minimum temperature for energy delivery is approximately 20°C.

tem parameter ranges used to compile the f-chart results:

ector transmissivity-absorbtivity, $(\tau\alpha)_n$	0.6-0.9
ector heat removal factor-area, $F_R A_c$	5-120 m^2
ector heat loss coefficient, U_c	2.1-8.3 W/m^2-°C
ector tilt angle, β	30°-90°
rall building loss coefficient, $\overline{UA}$	83-667 W/°C

gin with:

Monthly heating load; for information regarding water
heating and building heating loads refer to page 65.
**Monthly solar radiation totals for site-specific collector-
plane**; can be obtained from internet sources, page 83, or
for simple geometries, computed with the procedure
beginning on page 32.

Collector performance parameters $F_R \tau_s \alpha_s$ and $F_R U_c$;

can be obtained from the manufacturer or representative
values for non-concentrating collectors are given on page
46.
Solar system design parameters; this includes the
collector area, working fluid, fluid flow rate per unit area of
collector, storage capacity, and heat exchanger
performance. The standard configurations assumed with
this method are shown in Figs. f-1 and f-2 for water and air
based systems respectively.

ocedure begins on page 68.

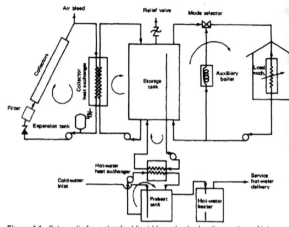

Figure f-1. Schematic for a standard liquid-based solar heating system. Note: cer[tain] deviations from this configuration can be handled by the f-chart method.

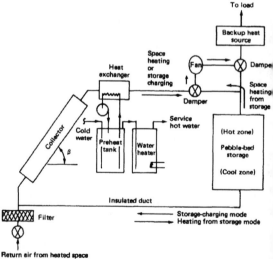

Figure f-2. Schematic for a standard air-based solar heating system. Note: certai[n] deviations from this configuration can be handled by the f-chart method.

ocedure:

Compute the loss parameter, P_L:

$$P_L = \frac{A_c \, F_{hx} \, F_R \, U_c \, \Delta t \left(T_R - \overline{T}_a \right)}{L}$$ Where A_c is the net collector

area (m^2), F_{hx} is the collector loop heat-exchanger factor (= 1 if no heat exchanger), Δt is the number of seconds per month, T_R is a reference temperature of 100° C, $\overline{T}_a$ is the monthly average ambient temperature (°C), and L is the total monthly heating load (J/month).

Compute the solar parameter, P_S:

$$P_s = \frac{A_c \, \overline{I}_c \, F_{hx} \, F_R \left(\tau\alpha \right)_n}{L} \left(\frac{F_R \overline{\left(\tau\alpha \right)}}{F_R \left(\tau\alpha \right)_n} \right)$$ Where $\overline{I}_c$ is the

total monthly collector-plane insolation (J/m^2-month) and

$$\left(\frac{F_R \overline{\left(\tau\alpha \right)}}{F_R \left(\tau\alpha \right)_n} \right) = 0.95$$ for collectors tilted within +/- 20° of the local latitude.

Compute modified parameters based on deviations from standard systems. See Table f-1 for liquid based systems, Table f-2 for air based systems and/or the modification for water heating-only (below).

ater-heating-only loss parameter modification:

$$= \frac{A_c \, F_{hx} \, F_R \, U_c \, \Delta t \left(11.6 + 1.18 T_{w,o} + 3.86 T_{w,i} - 2.32 \overline{T}_a \right)}{L}$$

Compute the solar supplied fraction, f_s, of the monthly heating load. Read f_s from the appropriate figure, (Fig. f-3 liquid/Fig. f-4 air) or use the associated expressions (below):

quid based systems:

$$= 1.029 P_s - 0.065 P_L - 0.245 P_s^2 + 0.0018 P_L^2 + 0.0215 P_s^3$$

· based systems:

$$= 1.040 P_s - 0.065 P_L - 0.159 P_s^2 + 0.00187 P_L^2 - 0.0095 P_s^3$$

·r the conditions:

$$\le P_s \le 3.0; \ 0 \le P_L \le 18.0; \ 0 \le f_s \le 1.0 \ \text{and} \ \begin{array}{l} P_s > P_L/12 \ \text{Liquid} \\ P_s > 0.07 P_L \ \text{Air} \end{array}$$

Table f-1. Liquid based system nominal values and modifying groups.

Parameter	Nominal value	Modified parameter[b]	
Flow rate[c] $\dfrac{(\dot{m}c_p)_f}{A_c}$	0.0128 liters H_2O equivalent/ sec · m_c^2	$P_L = P_{L,\text{nom}} \dfrac{F_{hc}F_R}{(F_{hc}F_R)_{\text{nom}}}$	(5.2
		$P_s = P_{s,\text{nom}} \dfrac{F_{hc}F_R}{(F_{hc}F_R)_{\text{nom}}}$	(5.2
Storage volume (water) $V_s = \left(\dfrac{M}{\rho A_c}\right)_s$	75 liters H_2O/m_c^2	$P_L = P_{L,\text{nom}} \left(\dfrac{V_s}{75}\right)^{-0.25}$	(5.2
Load heat exchanger[d] $\dfrac{\epsilon_L(\dot{m}c_p)_W}{Q_L}$	2.0	$P_s = P_{s,\text{nom}} \left\{ 0.393 + 0.651\exp\left[-0.139 \dfrac{Q_L}{\epsilon_L(\dot{m}c_p)_{\text{mn}}} \right.\right.$	(5.2

[c]Table prepared from data and equations presented in [2,8]

[b]Multiply basic definition of P_s and P_L in points 1 and 2 by factor for nonnominal group value. $(F_{hc}F_R)_{\text{nom}}$ refers to values of $F_{hc}F_R$ at collector rating or test conditions.

[c]In liquid systems the correction for flow rate is small and can usually be ignored if variation is more than 50 percent below the nominal value.

[d]$(\dot{m}c_p)_{\text{mn}}$ is the minimum fluid capacitance rate, usually that of air for the load heat exchanger; Q_L the heat load per unit temperature difference between inside and outside of the building.

Table f-2. Air based system nominal values and modifying factors.

Parameter	Nominal value[b]	Loss parameter multiplier
Storage capacity V_s	0.25 m^3/m_c^2	$\left(\dfrac{0.25}{V_s}\right)^{0.3}$
Fluid volumetric flow rate Q_c	10.1 liters/sec · m_c^2	$\left(\dfrac{Q_c}{10.1}\right)^{0.28}$

[a]Adapted from [2,8]

[b]Based on net collector area; fluid volume at standard atmosphere conditions.

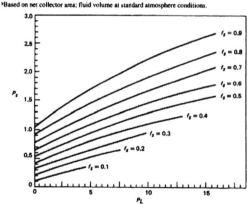

Figure f-3. f-chart for liquid based solar heating systems, [8].

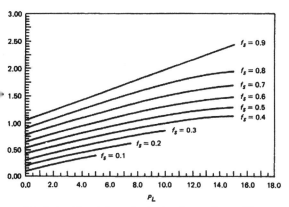

ure f-4. f-chart for air based solar heating systems, [8].

aylighting

neral design guidelines, from [7]

Orient building to maximize daylighting. Long axis running east-west preferred.

Maximize south glazing, minimize east and west facing glass. A south-facing aperture is the only orientation that, on an annual basis, balances typical thermal needs and lighting requirements with available radiation.

Optimally size overhangs on south-facing glazing to harvest daylighting, reduce summer heat gain, and permit the passive collection of solar thermal in winter.

Select the right glazing. Where windows are used specifically for daylighting, clear glass has an advantage over glazing with a low-E coating due to the coating's typical 10% to 30% reduction in visible light transmission.

Eliminate direct beam radiation. Use baffles to block direct beam radiation, diffuse light, and reduce glare.

Account for shading from adjacent buildings and trees and consider the reflectance from adjacent surfaces.

Use light colored roofing in front of monitors and select light colors for interior finishes to reflect in additional light and enhance distribution throughout the room

Daylighting Comparison

View Windows | High Sidelight | High Sidelight with Light-shelf | Wall Wash Toplightin

Central Toplighting | Patterned/Linear Toplighting | Tubular Skylights

Selection Criteria for Daylighting Strategies.

Design Criteria	View Windows (DL 1)	High Sidelight w/ Light Shelf (DL2 & DL3)	Wall Wash Toplighting (DL4)	Central & Patterned Toplighting (DL5 & DL6)	Linear Toplighting (DL 7)	Tubular Skylight (DL8)
Uniform Light Distribution	OO	●/O	●	●●	●	O
Low Glare	O	●	●	●	●	●/O
Reduced Energy Costs	O	●	●	●●	●	●
Cost Effectiveness	●	●	●	●	●	●●
Safety/Security Concerns	O	●/O	●	●	●	●●
Low Maintenance	O	●	●	●	●	●

●● Extremely good application ● Good application O Poor application OO Extremely poor application ● Depends
space layout and number and distribution of daylight apertures ●/O Mixed benefits

Reprinted from

Note: The configuration of the view window is with no controls, daylighting performance can be improved with appropriate measures, overhangs, shades, etc. Additionally, the benefit of providing a direct visual connection to the outdoors is no considered in this comparison.

Daylighting Design

Lumen method for estimating workplane illuminance level with sidelighting and skylighting.

Sidelighting with vertical windows:

The lumen method computes work plane illuminance levels for five reference points for the standard geometry shown in the figure d-1, next page.

delighting with vertical windows:

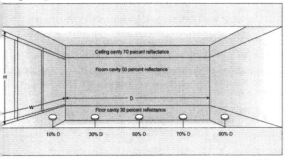

Figure d-1. Location of illumination points within the room (along the centerline of window) determined by lumen method of delighting, [6].

Determine the total sky illuminance entering the window, E_{sw}. Compute the solar altitude and azimuth angle for the desired latitude, date, and time of day, page 6. Compute the sun-window normal azimuth angle difference

using: $a_{sw} = |a_s - a_w|$ Using the figures for vertical

surfaces on page 74, determine the direct sun illuminance, $E_{v,sun}$, and the direct sky illuminance, $E_{v,sky}$, for the appropriate sky conditions. Then: $E_{sw} = E_{v,sun} + E_{v,sky}$.
Determine the reflected ground illuminance entering the window, E_{gw}. Read values of direct sun and sky illuminance for a horizontal surface, $E_{h,sun}$ and $E_{h,sky}$, page 74. For uniformly reflective ground surfaces extending from the window outward to the horizon, the illuminance on the window from ground reflection, E_{gw}, can be determined

with: $E_{gw} = \rho \left(E_{h,sun} + E_{h,sky} \right) / 2$ Where typical

reflectivity values, ρ, come from the table on page 36. Determine the net window transmittance, τ.

$\tau = TR_aT_cLLF$ where T is the glazing transmittance,

table d-1, R_A is the ratio of net to gross window areas and T_C is the transmittance of any light-controlling devices. Light loss factor values, LLF, can be taken from table d-2.

72

Sidelighting with vertical windows:

4. Compute the work plane illuminances for the geometry shown on page 72. $E_{TWP} = \tau \left(E_{sw} CU_{sky} + E_{gw} CU_g \right)$ where the coefficients of utilization for the sky component CU_{sky}, are taken from Table d-3, page 75-76, a-e dependi on the ratio of E_{vsky}/E_{hsky}. CU_g values are read from table d 3, f.

Table d-1. Glass transmittances

Glass	Thickness (in.)	τ
Clear	$\frac{1}{8}$	.89
Clear	$\frac{3}{16}$	.88
Clear	$\frac{1}{4}$	.87
Clear	$\frac{5}{16}$	.86
Grey	$\frac{1}{8}$	.61
Grey	$\frac{3}{16}$	.51
Grey	$\frac{1}{4}$	.44
Grey	$\frac{5}{16}$	.35
Bronze	$\frac{1}{8}$	.68
Bronze	$\frac{3}{16}$	.59
Bronze	$\frac{1}{4}$	.52
Bronze	$\frac{5}{16}$	.44
Thermopane	$\frac{1}{8}$	.80
Thermopane	$\frac{3}{16}$	.79
Thermopane	$\frac{1}{4}$	.77

Murdoch [11]

Table d-2. Light loss factors, [6].

Locations	Light Loss Factor Glazing Position		
	Vertical	Sloped	Horizontal
Clean Areas	0.9	0.8	0.7
Industrial Areas	0.8	0.7	0.6
Very Dirty Areas	0.7	0.6	0.5

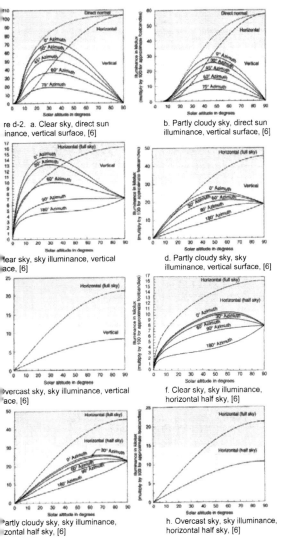

re d-2. a. Clear sky, direct sun
inance, vertical surface, [6]

b. Partly cloudy sky, direct sun
illuminance, vertical surface, [6]

lear sky, sky illuminance, vertical
ace, [6]

d. Partly cloudy sky, sky
illuminance, vertical surface, [6]

Overcast sky, sky illuminance, vertical
ace, [6]

f. Clear sky, sky illuminance,
horizontal half sky, [6]

Partly cloudy sky, sky illuminance,
zontal half sky, [6]

h. Overcast sky, sky illuminance,
horizontal half sky, [6]

74

Sidelighting with vertical windows:

Table d-3. Coefficients of utilization for lumen method of sidelighting (window without blinds), [6].

a. CUsky for (Ev/Eh)sky = 0.75

Room Depth/Window Height	Percent D*	Window Width/Window Height					
		.5	1	2	3	4	6	8	Infinite

Room Depth/Window Height	Percent D*	.5	1	2	3	4	6	8	Infinite
1	10	.624	.684	.820	.873	.875	.879	.880	.883
	30	.547	.611	.777	.799	.793	.793	.798	.810
	50	.355	.526	.659	.666	.660	.669	.670	.672
	70	.243	.365	.538	.548	.548	.444	.446	.447
	90	.185	.304	.418	.451	.464	.444	.446	.447
2	10	.567	.781	.809	.812	.813	.816	.818	.824
	30	.298	.418	.519	.544	.561	.556	.557	.563
	50	.121	.169	.175	.203	.206	.338	.344	.345
	70	.068	.116	.170	.201	.215	.259	.284	.289
	90	.050	.083	.127	.151	.164	.167	.171	.172
3	10	.522	.681	.709	.746	.747	.747	.749	.615
	30	.138	.232	.330	.350	.360	.365	.366	.373
	50	.053	.092	.139	.163	.174	.183	.182	.187
	70	.031	.053	.081	.097	.106	.116	.116	.119
	90	.025	.041	.063	.074	.082	.090	.090	.092
4	10	.405	.578	.670	.873	.875	.674	.707	.558
	30	.075	.134	.197	.224	.235	.243	.243	.153
	50	.011	.028	.049	.044	.055	.058	.056	.040
	70	.009	.018	.027	.032	.035	.040	.041	.040
	90	.016	.026	.040	.048	.053	.058	.061	.041
6	10	.147	.257	.352	.390	.391	.382	.523	.452
	30	.027	.049	.066	.090	.054	.070	.070	.063
	50	.011	.019	.028	.034	.037	.042	.044	.040
	70	.006	.011	.017	.022	.024	.019	.033	.034
	90	.005	.008	.013	.016	.019	.021	.025	.025
10	10	.092	.182	.248	.276	.284	.290	.291	.395
	30	.014	.026	.039	.044	.046	.044	.051	.087
	50	.005	.010	.015	.018	.020	.022	.024	.032
	70	.003	.007	.010	.012	.014	.016	.018	.017
	90	.003	.006	.011	.013	.015	.016	.017	.024

*Percent D is the relative distance from the window to the opposite wall.

b. CUsky for (Ev/Eh)sky = 1.00

Room Depth/Window Height	Percent D*	Window Width/Window Height							
		.5	1	2	3	4	6	8	Infinite
1	10	.671	.704	.711	.715	.717	.726	.728	.729
	30	.458	.595	.654	.668	.598	.599	.609	.665
	50	.313	.452	.560	.589	.590	.607	.609	.610
	70	.227	.362	.478	.515	.527	.481	.468	.534
	90	.196	.308	.424	.465	.481	.468	.471	.472
2	10	.545	.635	.661	.661	.665	.665	.665	.672
	30	.239	.367	.459	.484	.491	.499	.500	.506
	50	.073	.099	.203	.192	.209	.344	.366	.366
	70	.058	.101	.155	.188	.207	.243	.241	.287
	90	.000	.061	.098	.110	.123	.207	.216	.223
3	10	.431	.561	.607	.747	.748	.615	.615	.631
	30	.103	.223	.307	.337	.348	.357	.367	.366
	50	.058	.103	.155	.183	.197	.211	.213	.218
	70	.057	.061	.094	.118	.132	.147	.150	.154
	90	.000	.049	.074	.098	.110	.122	.126	.129
4	10	.339	.482	.540	.560	.563	.565	.565	.593
	30	.078	.139	.234	.247	.258	.268	.260	.272
	50	.022	.039	.060	.090	.114	.139	.143	.150
	70	.011	.021	.033	.040	.061	.064	.073	.089
	90	.019	.032	.052	.061	.070	.061	.070	.089
6	10	.211	.343	.433	.453	.461	.468	.461	.518
	30	.033	.059	.035	.123	.057	.064	.075	.197
	50	.015	.029	.047	.057	.063	.073	.077	.086
	70	.011	.021	.031	.040	.044	.051	.054	.060
	90	.010	.019	.029	.034	.038	.044	.048	.052
10	10	.135	.238	.310	.336	.342	.382	.290	.452
	30	.016	.034	.058	.072	.088	.090	.064	.118
	50	.006	.017	.025	.017	.020	.045	.048	.059
	70	.008	.011	.018	.021	.024	.031	.046	.045
	90	.003	.008	.014	.015	.018	.023	.031	.038
.10	10	.060	.165	.217	.272	.283	.290	.291	.395
	30	.006	.024	.034	.039	.045	.052	.087	.087
	50	.005	.013	.022	.024	.033	.023	.025	.044
	70	.004	.009	.015	.018	.015	.016	.033	.033
	90	.003	.008	.014	.016	.018	.022	.025	.022

*Percent D is the relative distance from the window to the opposite wall.

c. CUsky for (Ev/Eh)sky = 1.25

Room Depth/Window Height	Percent D*	Window Width/Window Height							
		.5	1	2	3	4	6	8	Infinite
1	10	.579	.607	.614	.618	.621	.630	.634	.635
	30	.405	.525	.590	.604	.596	.612	.614	.615
	50	.287	.423	.519	.547	.560	.566	.571	.573
	70	.212	.347	.461	.473	.491	.516	.525	.528
	90	.168	.307	.428	.461	.491	.488	.466	.487
2	10	.472	.548	.568	.569	.570	.574	.575	.581
	30	.221	.337	.422	.447	.466	.465	.467	.472
	50	.082	.155	.203	.211	.238	.281	.287	.361
	70	.057	.098	.174	.211	.222	.281	.284	.290
	90	.050	.080	.112	.174	.181	.227	.251	.253
3	10	.377	.488	.527	.535	.534	.535	.538	.549
	30	.130	.217	.298	.328	.341	.352	.353	.382
	50	.082	.110	.185	.195	.195	.225	.231	.237
	70	.040	.070	.109	.132	.147	.166	.171	.175
	90	.057	.060	.112	.127	.142	.146	.140	.152
4	10	.300	.424	.484	.494	.497	.499	.499	.524
	30	.069	.143	.206	.240	.255	.267	.266	.283
	50	.035	.088	.094	.135	.124	.158	.163	.188
	70	.026	.047	.073	.085	.094	.109	.115	.106
	90	.021	.031	.058	.085	.074	.084	.092	.103
6	10	.183	.314	.395	.415	.420	.423	.423	.476
	30	.024	.059	.113	.136	.155	.146	.165	.186
	50	.017	.033	.053	.063	.071	.074	.085	.100
	70	.010	.018	.031	.043	.049	.056	.061	.080
	90	.010	.021	.031	.037	.043	.049	.053	.066
10	10	.129	.228	.300	.337	.348	.361	.363	.433
	30	.019	.039	.062	.082	.082	.104	.109	.134
	50	.008	.021	.032	.038	.042	.052	.059	.060
	70	.006	.013	.020	.025	.028	.035	.035	.049
	90	.004	.009	.015	.018	.020	.026	.029	.043
	10	.098	.194	.241	.276	.290	.291	.291	.396
	30	.011	.034	.044	.052	.054	.066	.087	.087
	50	.009	.024	.034	.038	.040	.033	.038	.103
	70	.008	.012	.022	.028	.028	.038	.038	.060
	90	.004	.010	.017	.018	.018	.023	.025	.038

*Percent D is the relative distance from the window to the opposite wall.

delighting with vertical windows:

ble d-3. Coefficients of utilization for lumen method of
delighting (window without blinds), [6].

f. CUlg (ground reflected component)

*Percent D is the relative distance from the window to the opposite wall.

e. CUsky for (EvEh)sky = 1.75

*Percent D is the relative distance from the window to the opposite wall.

d. CUsky for (EvEh)sky = 1.50

*Percent D is the relative distance from the window to the opposite wall.

Lumen method for skylighting:

1. **Determine the total sky illuminance entering the skylight.** Compute the solar altitude angle for the desire latitude, date, and time of day, page 6. Determine the tot horizontal skylight illuminance, E_H, (direct sun plus sky illuminance) using the appropriate horizontal surface data Figure d-2, page 74.

2. **Determine the net skylight transmittance, T_n.** The flat plate transmittances, T_F, for several plastic materials are provided in Table d-3 below. For domed skylights, the effective dome transmittance, T_D, can be computed using

   $$T_D = 1.25 T_F \left(1.18 - 0.416 T_F \right)$$ For double-domed

   skylights, the individual transmittances can be combined

   using: $$T_D = \frac{T_{D1} T_{D2}}{T_{D1} T_{D2} - T_{D1} T_{D2}}$$ Also determine the light lo

   factor, LLF from Table d-2, page 73. Compute the well

   cavity ratio for the skylight well with: $$WCR = \frac{5h \left(w + \right.}{wl}$$

   Using Figure d-3, page 78, determine the skylight well efficiency, N_w. Compute the net skylight transmittance, T

   using: $$T_n = T_D N_w R_A T_C LLF$$ where R_A is the ratio of net t

   gross skylight areas and T_C is the transmittance of any light-controlling devices.

Table d-3. Flat-plate plastic material transmittance for skyligh
Source: Murdoch[11]

Type	Thickness (in.)	Transmittan
Transparent	$\frac{1}{8} - \frac{1}{16}$	.92
Dense translucent	$\frac{1}{8}$	.32
Dense translucent	$\frac{3}{16}$	.24
Medium translucent	$\frac{1}{8}$	.56
Medium translucent	$\frac{3}{16}$	.52
Light translucent	$\frac{1}{8}$	.72
Light translucent	$\frac{3}{16}$	.68

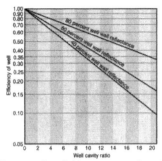

ure d-3. Efficiency of well, N_w, versus well cavity ratio, [6].

Compute the room coefficient of utilization, CU. For office and warehouse interiors CU can be estimated with:

$$CU = \frac{1}{1 + A\left(RCR\right)^{B}} \quad \text{for RCR} < 8 \quad \text{where}$$

A=0.0288, B=1.560 for offices (typical reflectance: ceiling 0.75, wall 0.5, floor 0.3) and A=0.0995, B=1.087 for warehouses (typical reflectance: ceiling 0.5, wall 0.3, floor 0.2). The room cavity ratio is given by

$$RCR = \frac{5h_c\left(l + w\right)}{lw} \quad \text{with } h_c \text{ being the ceiling height}$$

above the work plane and l and w being the room length and width, respectively.

Compute the illuminance at the work plane, E_{TWP}:

$$E_{TWP} = E_H T_n \left(CU\right)\left(\frac{A_T}{A_{WP}}\right) \quad \text{:where } A_T \text{ is the total gross}$$

area of the skylights (number of skylights times the skylight gross area), and A_{WP} is the work plane area (typically room length times width). Note that it is possible to fix the E_{TWP} at some desired value and determine the skylight area required, however, due to their dependence on A_T, the factors N_w and R_A (step 2) should be recalculated.

Solar Dryer Configurations

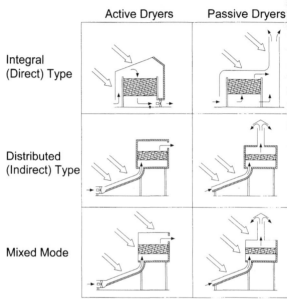

Active Dryers · Passive Dryers

Integral (Direct) Type

Distributed (Indirect) Type

Mixed Mode

Adapted from

Safe storage moisture for aerated good quality grain, [15].

Grain	Maximum safe moisture content
Shelled corn and sorghum	
To be sold as #2 grain or equiv. by spring	15.5%
To be stored up to 1 year	14%
To be stored more than 1 year	13%
Soybeans	
To be sold by spring	14%
To be stored up to 1 year	12%
Wheat	13%
Small grain (oats, barley, etc.)	13%
Sunflowers	9%

Life Cycle Costing

e Cycle Cost (LCC) is an economic measure that reflects the
nefits accrued by solar usage throughout the lifetime of a
lar-powered system. This is opposed to a simple initial cost
mparison, which does not include lifetime energy consumption
d, therefore, typically favors non-renewable alternatives.

C is the sum of the present worth (pw) values of all of the
penses associated with the system over its expected lifetime:

$$C = C + M_{pw} + E_{pw} + R_{pw} - S_{pw}$$

ere:

capital cost of a project which includes the initial
capital expense for equipment, the system design,
engineering, and installation. This cost is always
considered as a single payment occurring in the initial
year of the project, regardless of how the project is
financed.

maintenance is the sum of all yearly scheduled
operation and maintenance (O&M) costs. Fuel or
equipment replacement costs are not included.

energy cost is the sum of the yearly fuel or electricity
cost for the system.

replacement cost is the sum of all repair and
equipment replacement cost anticipated over the life
of the system.

salvage value of a system is its net worth in the final
year of the life-cycle period. It is common practice to
assign a salvage value of 20% of original cost for
mechanical equipment that can be moved.

mputation of the present worth of future expenditures.

The single present worth (P) of a future sum of money (F)
in a given year (N) at a given investment rate (D) and
inflation rate (i) is:

$$P = FX^N \quad \text{with} \quad X = \left(\frac{1+i}{1+D} \right)$$

2. The uniform present worth (P) of an annual sum (A) received over a period of years (N) at a given investmen rate (D) and inflation rate (i) is:

$$P = A\left(1 - X^N\right)/\left(X^{-1} - 1\right)$$

Example: Compare the life cycle cost for a photovoltaic pow supply with battery storage and a gasoline engine-generator alternative (systems specified below).

Given:
Life cycle period:	20 years
Investment rate:	7%
General inflation:	3%
Fuel inflation:	4%

Solution: Begin with system designs for each alternative, making sure that each system provides equivalent performan for example: power output, reliability, lifetime, etc. Using a detailed cost estimate (sample below), the present worth of e component is determined. For future one-time expenditures, (e.g. battery replacement, generator rebuild) point 1 above ca be used. For costs repeating on an annual basis (e.g. maintenance, generator fuel) point 2 is used. Note the use of separate fuel inflation rate for fuel expenses. The present wo values are then summed (less salvage) to give the LCC.

PV System

Item	Initial Cost ($)	Present Worth ($)
1. Capital		
Array	2500	2500
Controller	300	300
Batteries	900	900
Installation	700	700
2. Maintenance		
Annual Inspection	75	1030
(per year)		
3. Energy		
None	-	-

continued

Item	Initial Cost ($)	Present Worth ($)
4. Replacement		
Battery bank @ yr. 5	900	744
Battery bank @ yr. 10	900	615
Battery bank @ yr. 15	900	509
5. Salvage		
20% of original equipment cost	740	(346)
LCC: (Items 1 + 2 + 3 + 4 - 5)		$6,952

Engine-generator System

Item	Initial Cost ($)	Present Worth ($)
Capital		
Generator	400	400
Installation	300	300
Maintenance		
Tune-up (per year)	150	2060
Annual Inspection (per year)	75	1030
Energy		
Annual fuel cost (4% fuel inflation rate)	375	5640
Replacement		
Gen. rebuild @ yr. 5	250	207
Gen. rebuild @ yr. 10	250	171
Gen. rebuild @ yr. 15	250	142
Salvage		
20% of original equipment cost	80	(38)
LCC: (Items 1 + 2 + 3 + 4 - 5)		$9,912

References

Internet sources of data

Solar Resource Data

World radiation data center (WRDC) online archive, Russian Federal Service for Hydrometeorology and Environmental Monitoring; 1964-1993 data http://wrdc-mgo.nrel.gov/ 1994-present data http://wrdc.mgo.rssi.ru/

Surface meteorology and solar energy, National Aeronautics a Space Administration, USA; http://eosweb.larc.nasa.gov/sse

Solar radiation resource information, National Renewable Energy Laboratory, USA; http://rredc.nrel.gov/solar

Climatic Data

World climatic data, World Weather Information Service; http://www.worldweather.org

U. S. climate data, National Oceanic and Atmospheric Administration, USA; http://www.noaa.gov/climate.html

ted References

ASHRAE. 2001. *2001 ASHRAE Handbook, Fundamentals.* Atlanta: American Society of Heating, Refrigerating and Air-Conditioning Engineers, Inc.

Duffie, J. A. and Beckman, W. A. 1991. *Solar Engineering of Thermal Processes.* 2nd ed. New York: John Wiley & Sons.

Building Technologies Program, Lawrence Berkeley National Lab. 1997. *Tips for Daylighting with Windows.* Berkeley: Lawrence Berkeley National Lab.

Collares-Pereira and Rabl, A. 1979. Simple Procedure for Predicting Long Term Average Performance of Nonconcentrating and of Concentrating Solar Collectors. *Solar Energy,* Vol. 23, pg. 235-254.

Goswami, D. Y., Kreith, F., and Kreider, J. 2000. *Principles of .Solar Engineering,* 2nd ed. Philadelphia: Taylor & Francis.

IESNA. 2000. *The IESNA Lighting Handbook.* New York: Illumination Engineering Society of North America. Reprinted with permission from the IESNA Lighting Handbook, 9th Edition, courtesy of the Illuminating Engineering Society of North America.

Innovative Design, Inc. 2004. *Guide for Daylighting Schools.* Raleigh, NC: Innovative Design, Inc.

Klein, S. A. 1976. *A Design Procedure for Solar Heating Systems,* Ph.D. dissertation, Univ. of Wisconsin, Madison. For an approach similar to the f-chart for other solar-thermal systems operating above a minimum temperature above that for space-heating (~20°C), see Klein, S. A. and Beckman, W. A. 1977. A General Design Method for Closed Loop Solar Energy Systems. *Proc. 1977 ISES Meeting.*

Löf, G. O. G. and Tybout, R. A. 1972. Model for Optimizing Solar Heating Design. ASME Paper 72-WA/SOL-8.

Messenger, R. and Ventre, J. 2000. *Photovoltaic Systems Engineering.* Boca Raton, FL: CRC Press.

Murdoch, J. B. 1985. *Illumination Engineering: From Edison's Lamp to the Laser.* New York: Macmillan Publishing Co.

Norton, B. 1992. *Solar Energy Thermal Technology.* London: Springer-Verlag.

Post, H. N. and Risser, V. V. 1995. *Stand-Alone Photovoltaic Systems—A Handbook of Recommended Design Practices.* Albuquerque: Sandia National Lab. Report SAND-87-7023.

U.S. Department of Energy. 2002. *National Best Practices Manual for Building High Performance Schools.* Publication DOE/GO-102002-1610.

Midwest Plan Service. 1980. *Low Temperature and Solar Grain Drying Handbook.* Ames, IA: Iowa State University.

Units and Conversion Factors

Fundamental SI units

Quantity	Name of unit	Symbol
Length	Meter	m
Mass	Kilogram	kg
Time	Second	sec
Electric current	Ampere	A
Thermodynamic temperature	Kelvin	K
Luminous intensity	Candela	cd
Amount of a substance	Mole	mol

Derived SI units

Quantity	Name of unit	Symbol
Acceleration	Meters per second squared	m/sec^2
Area	Square meters	m^2
Density	Kilogram per cubic meter	kg/m^3
Dynamic viscosity	Newton-second per square meter	$N \cdot sec/m^2$
Force	Newton (= $1 kg \cdot m/sec^2$)	N
Frequency	Hertz	Hz
Kinematic viscosity	Square meter per second	m^2/sec
Plane angle	Radian	rad
Potential difference	Volt	V
Power	Watt (= 1 J/s)	W
Pressure	Pascal (= $1 N/m^2$)	Pa
Radiant intensity	Watts per steradian	W/sr
Solid angle	Steradian	sr
Specific heat	Joules per kilogram-Kelvin	$J/kg \cdot K$
Thermal conductivity	Watts per meter-Kelvin	$W/m \cdot K$
Velocity	Meters per second	m/sec
Volume	Cubic meter	m^3
Work, energy, heat	Joule (= $1 N \cdot m$)	J

Fundamental constants

Quantity	Symbol	Value
Avogadro constant	N	6.022169×10^{26} kmol^{-1}
Boltzmann constant	k	1.380622×10^{-23} J/K
First radiation constant	$C_1 = 2\pi hc^2$	3.741844×10^{-16} W $\cdot$ m^2
Gas constant	R	8.31434×10^3 J/kmol $\cdot$ K
Planck constant	h	6.626196×10^{-34} J $\cdot$ sec
Second radiation constant	$C_2 = hc/k$	1.438833×10^{-2} m $\cdot$ K
Speed of light in a vacuum	c	2.997925×10^8 m/sec
Stefan-Boltzmann constant	σ	5.66961×10^{-8} W/m$^2 \cdot$ K^4

sical quantity	Symbol	Conversion factor
a	A	1 ft^2 = 0.0929 m^2
		1 acre = 43,560 ft^2 = 4047 m^2
		1 hectare = 10,000 m^2
		1 square mile = 640 acres
sity	ρ	1 lb$_m$/ft^3 = 16.018 kg/m^3
t, energy, or work	Q or W	1 Btu = 1055.1 J
		1 kWh = 3.6 MJ
		1 Therm = 105.506 MJ
		1 cal = 4.186 J
		1 ft · lb$_f$ = 1.3558 J
ce	F	1 lb$_f$ = 4.448 N
t flow rate, Refrigeration	q	1 Btu/hr = 0.2931 W
		1 ton (refrigeration) = 3.517 kW
		1 Btu/sec = 1055.1 W
t flux	q/A	1 Btu/hr · ft^2 = 3.1525 W/m^2
t-transfer coefficient	h	1 Btu/hr · ft^2 · F = 5.678 W/m^2 · K
gth	L	1 ft = 0.3048 m
		1 in = 2.54 cm
		1 mi = 1.6093 km
ıs	m	1 lb$_m$ = 0.4536 kg
		1 ton = 2240 lbm
		1 tonne (metric) = 1000 kg
ıs flow rate	$\dot{m}$	1 lb$_m$/hr = 0.000126 kg/sec
›er	$\dot{W}$	1 hp = 745.7 W
		1 kW = 3415 Btu/hr
		1 ft · lb$_f$/sec = 1.3558 W
		1 Btu/hr = 0.293 W
sure	p	1 lb$_f$/in^2 (psi) = 6894.8 Pa (N/m^2)
		1 in. Hg = 3,386 Pa
		1 atm = 101,325 Pa (N/m^2) = 14.696 psi
iation	l	1 langley = 41,860 J/m^2
		1 langley/min = 697.4 W/m^2
cific heat capacity	c	1 Btu/lb$_m$ · °F = 4187 J/kg · K
rnal energy or enthalpy	e or h	1 Btu/lb$_m$ = 2326.0 J/kg
		1 cal/g = 4184 J/kg
ıperature	T	$T(°R) = (9/5)T(K)$
		$T(°F) = [T(°C)](9/5) + 32$
		$T(°F) = [T(K) - 273.15](9/5) + 32$
rmal conductivity	k	1 Btu/hr · ft · °F = 1.731 W/m · K
rmal resistance	R_{th}	1 hr · °F/Btu = 1.8958 K/W
›city	V	1 ft/sec = 0.3048 m/sec
		1 mi/hr = 0.44703 m/sec
:osity, dynamic	μ	1 lb$_m$/ft · sec = 1.488 N · sec/m^2
		1 cP = 0.00100 N · sec/m^2
:osity, kinematic	ν	1 ft^2/sec = 0.09029 m^2/sec
		1 ft^2/hr = 2.581 × 10^{-5} m^2/sec
ıme	V	1 ft^3 = 0.02832 m^3 = 28.32 liters
		1 barrel = 42 gal (U.S.)
		1 gal (U.S. liq.) = 3.785 liters
		1 gal (U.K.) = 4.546 liters
ımetric flow rate	$\dot{Q}$	1 ft^3/min (cfm) = 0.000472 m^3/sec
		1 gal/min (GPM) = 0.0631 l/sec

The History of ISES

1954	ISES has its origin in Phoenix, Arizona, USA. A group of indust[ ] financial and agricultural leaders establishes the "Association for App[ ] Solar Energy" (AFASE) as a non-profit organization.
1955	The first two important meetings are held in Tucson and Phoenix, U[ ] which attract more than 1000 scientists, engineers and governm[ ] officials from 36 different countries.
1956	The association establishes its first scientific publication "the su[ ] work".
1957	The first issue of "The Journal of Solar Energy, Science [ ] Engineering" is published.
1963	Dedicated solar scientists decide that radical changes are required[ ] the operation and goals of the society. Through the reorganiza[ ] within the framework of its original concept the name is changed to "[ ] Solar Energy Society". Accreditation of the society at the United Nati[ ] Economic & Social Council (ECOSOC).
1964	The name of the journal is changed to "Solar Energy – The Journa[ ] Solar Energy Science and Technology". Prof. Farrington Daniels is elected President and in his honor [ ] 'Farrington Daniels Award' is established in 1975. The award dign[ ] outstanding intellectual leadership in the field of renewable energy.
1970	Relocation of the international headquarters office to Melbou[ ] Australia. First international conference outside the USA is held [ ] Melbourne, Australia.
1971	The name of the society is changed to "International Solar Ene[ ] Society".
1976	The first issue of the ISES magazine "SunWorld" is published.
1979	ISES celebrates its Silver Jubilee at the Solar World Congress in Atla[ ] USA.
1980	The 'Achievement through Action Award' in memory of Christophe[ ] Weeks is set up. A substantial cash prize honors contributions [ ] practical use or new concepts.
1989	ISES introduces the "Section Sponsorship Programme" in w[ ] individuals and sections have the opportunity to sponsor sections f[ ] developing countries.
1992	Accepted by the United Nations as a non-governmental organiza[ ] (NGO) in consultative status, ISES actively participates in the Un[ ] Nations Conference on Environment and Development (UNCED), hel[ ] Rio de Janeiro, Brazil.
1994	ISES coordinates the second meeting of the United Nations Commiss[ ] on Sustainable Development (CSD) in New York to discuss issues[ ] sustainable development and the interlinkages to renewable ene[ ] technologies.
1995	The Headquarters office moves from Melbourne, Australia to Freib[ ] Germany. The Headquarters becomes a focal point for internatio[ ] projects.
2000	Launch of new official ISES magazine "Refocus" (Renewable Ene[ ] Focus).

02	The first woman elected as President. Prof. Anne Grete Hestnes takes over office on January 1st, 2002.
	ISES participates in the World Summit on Sustainable Development (WSSD) in Johannesburg, South Africa.
05	Established on December 24, 1954 the society celebrates its Golden Jubilee at the Solar World Congress in Orlando, Florida, USA in August 2005.